Ronan Powell

Key Stage Three

Science

There's a lot to learn in KS3 Science... it's enough to steam up anyone's safety goggles. Not to worry — this brilliant all-in-one CGP book will help you see through the mist.

It's packed with clear notes and diagrams that explain every topic, plus plenty of practice questions and mixed-topic tests to make sure you've understood everything.

And once you've worked through that lot, there's a practice exam at the end of the book to *really* put your Science knowledge to the test!

How to access your free Online Edition

This book includes a free Online Edition to read on your PC, Mac or tablet. You'll just need to go to **cgpbooks.co.uk/extras** and enter this code:

3042 7570 4489 7706

By the way, this code only works for one person. If somebody else has used this book before you, they might have already claimed the Online Edition.

Complete

Study & Practice

Everything you need for the whole course!

Contents

Section One — Cells and Respiration

Section Two — Humans as Organisms

Section Three — Plants and Ecosystems

Section Four — Inheritance, Variation and Survival

Section Five — Classifying Materials

Section Six — Chemical Changes

Section Seven — The Earth and The Atmosphere

Section Eight — Energy and Matter

Section Nine — Forces and Motion

Section Ten — Waves

Section Eleven — Electricity and Magnetism

Section Twelve — The Earth and Beyond

Section Thirteen — Exam Practice

Published by CGP

From original material by Richard Parsons and Paddy Gannon.

Editors:
Mary Falkner, Christopher Lindle, Duncan Lindsay, Frances Rooney, Ethan Starmer-Jones and Charlotte Whiteley

Contributors:
Josephine Horlock and Lucy Muncaster

ISBN: 978 1 84146 385 8

Printed by Elanders Ltd, Newcastle upon Tyne.
Clipart from Corel®

The Microscope

A microscope is used for looking at objects that are too small to see with the naked eye. The lenses in the microscope magnify objects (make them look bigger) so that you can see them.

Learn the Different Parts of a Microscope

Here are some of the main parts of a light microscope — make sure you can identify them.

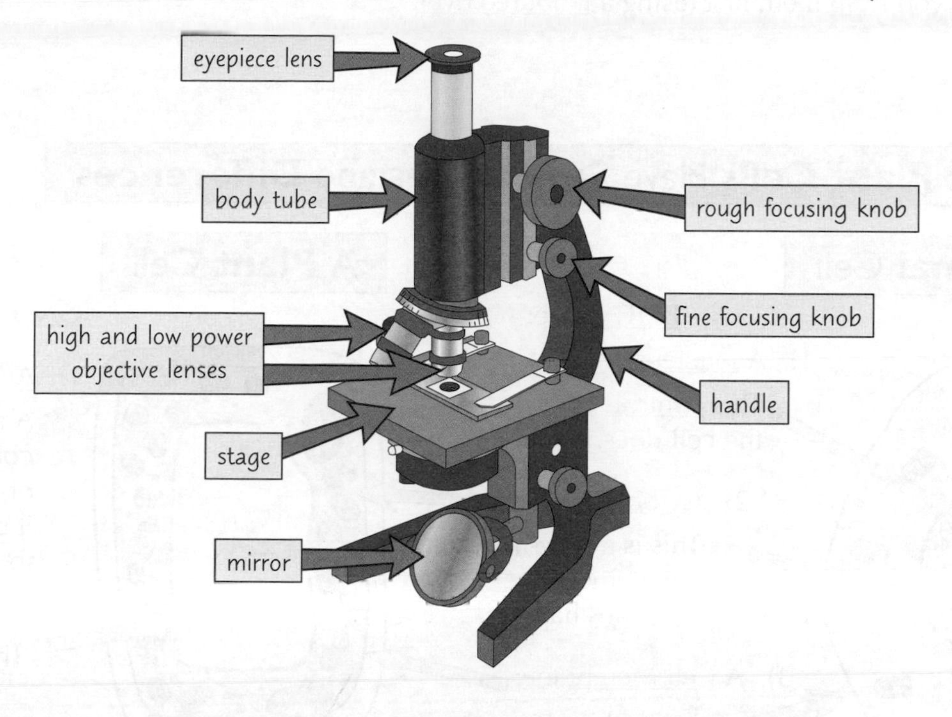

Follow These Easy Steps to Using a Light Microscope

1) Carry your microscope by the handle.
2) Place it near a lamp or a window, and angle the mirror so light shines up through the hole in the stage.
3) Clip a slide onto the stage. The slide should have the object(s) you want to look at stuck to it.
4) Select the lowest powered objective lens.

Don't reflect direct sunlight into the microscope — it could damage your eyes.

5) Turn the rough focusing knob to move the objective lens down to just above the slide.
6) Look down the eyepiece lens and adjust the focus using the fine focusing knob.
7) Keep adjusting until you get a clear image of whatever's on the slide.

DON'T BREAK THE SLIDE

Always turn the fine focusing knob so that the objective lens is moving away from the slide — so the lens and slide don't crash together.

8) If you need to see the slide with greater magnification, switch to a higher powered objective lens (a longer one).
9) Now refocus the microscope (repeat steps 5 to 7).

Microscopes are great for looking at cells

A microscope lets you see all the tiny building blocks (called cells) that make up living things. Choosing the correct equipment and using it properly and safely is a key part of being a scientist.

Cells

Living Things are Made of Cells

1) Another word for a living thing is an organism.
   All organisms are made up of tiny building blocks known as cells.
2) Cells can be seen through a microscope (see previous page) — but it helps if you stain them first (using a coloured dye).

Animal and Plant Cells Have Similarities and Differences

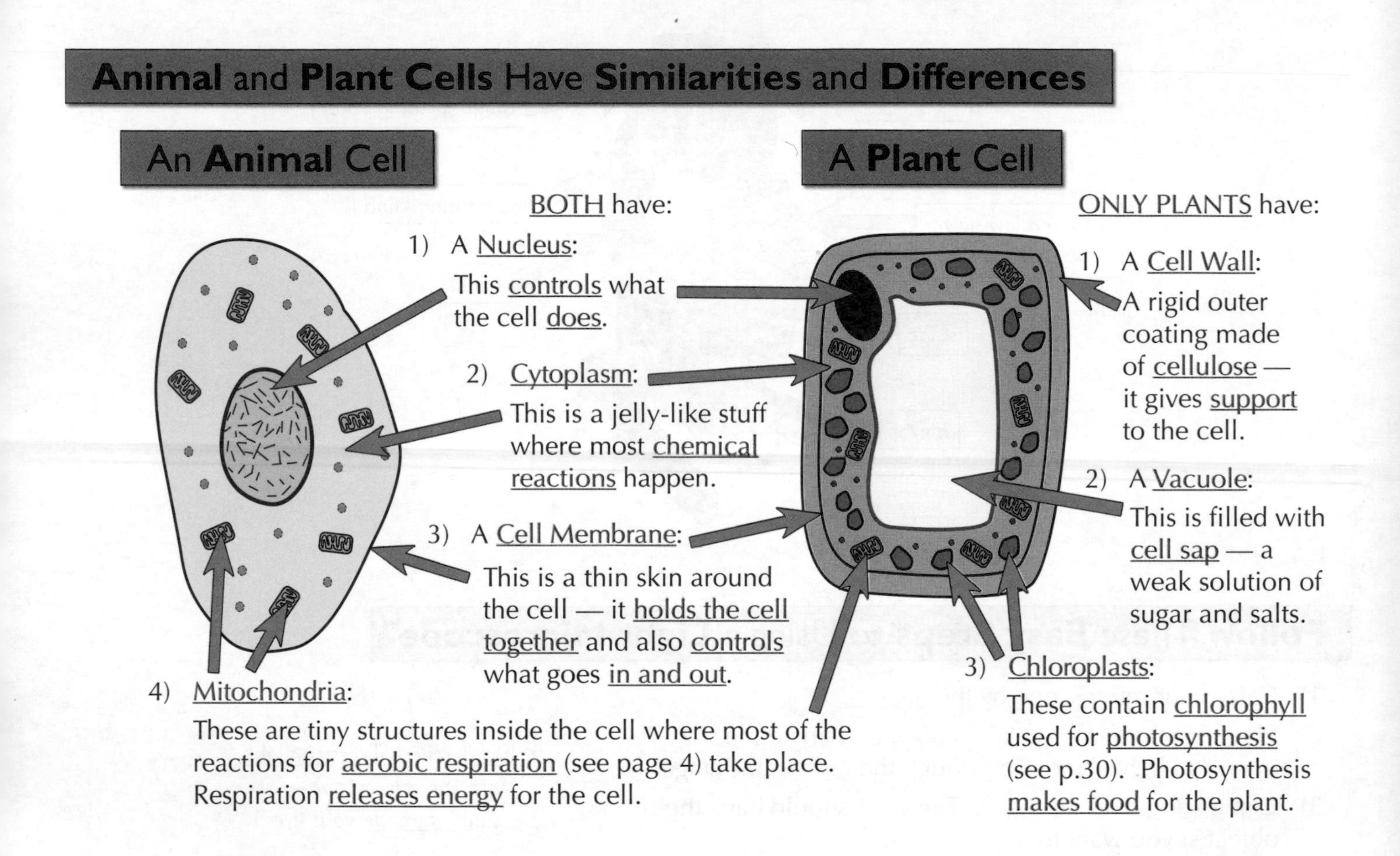

Some Living Things are Unicellular

1) Animals and plants are made up of lots of cells. They're multicellular organisms.
2) But many living things are made up of only one cell — these are called unicellular organisms. Unicellular organisms have adaptations to help them survive in the environment they live in, e.g.

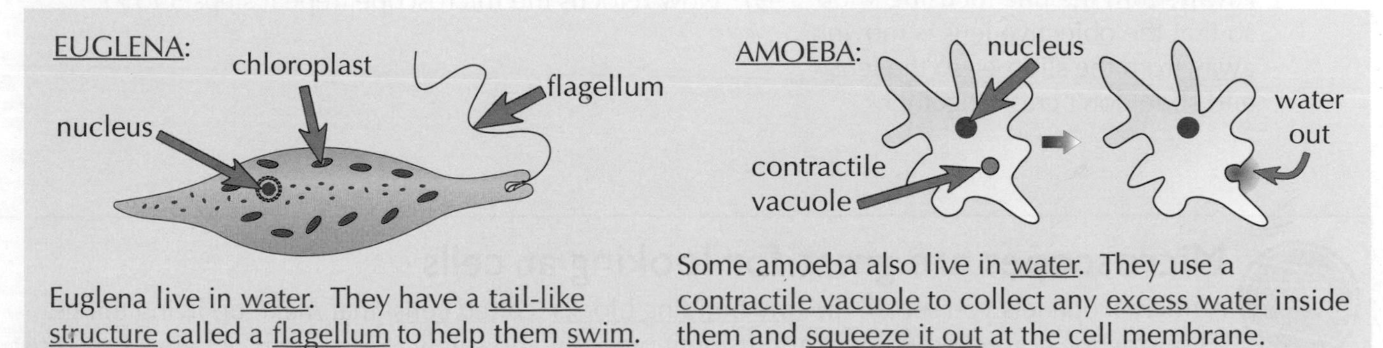

Euglena live in water. They have a tail-like structure called a flagellum to help them swim.

Some amoeba also live in water. They use a contractile vacuole to collect any excess water inside them and squeeze it out at the cell membrane.

Cell Organisation

Learn How Cells are Organised

In organisms with lots of cells (like animals and plants), the cells are organised into groups. Here's how:

A group of similar cells come together to make a tissue.
A group of different tissues work together to make an organ.
A group of organs work together to make an organ system.
A multicellular organism is usually made up of several organ systems.

Here's a rather jolly example from a plant.
Don't forget that the sequence applies just as well to animals.

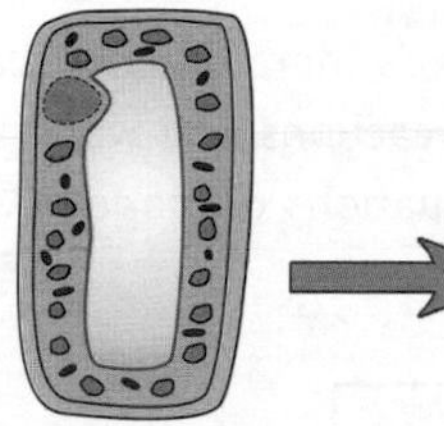

palisade CELLS...

...make up palisade TISSUE...

A palisade cell is just the name for a particular type of plant cell.

...which, with other tissues, makes up a leaf (an ORGAN)...

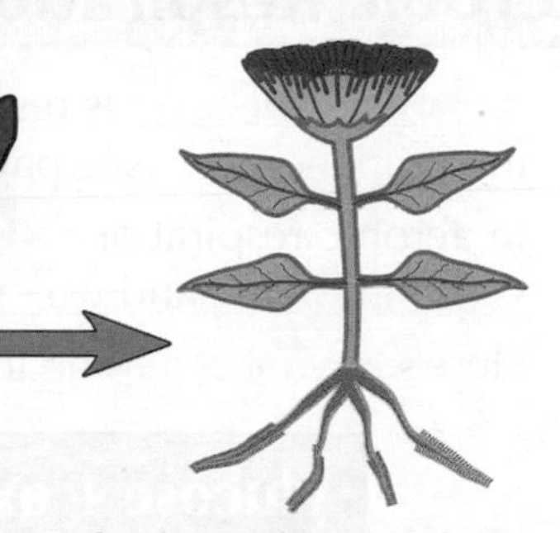

...which, with more leaves and other organs, makes up the shoot system (an ORGAN SYSTEM). Different organ systems make up a full plant (an ORGANISM).

Stuff Moves Into and Out of Cells by Diffusion

1) Cells need things like glucose (a sugar) and oxygen to survive. They also need to get rid of waste products, like carbon dioxide.
2) These materials all move into or out of cells by a process called diffusion.
3) Diffusion is where a substance moves from an area of high concentration (where there's lots of it) to an area of low concentration (where there's less of it) — just like glucose in this diagram...

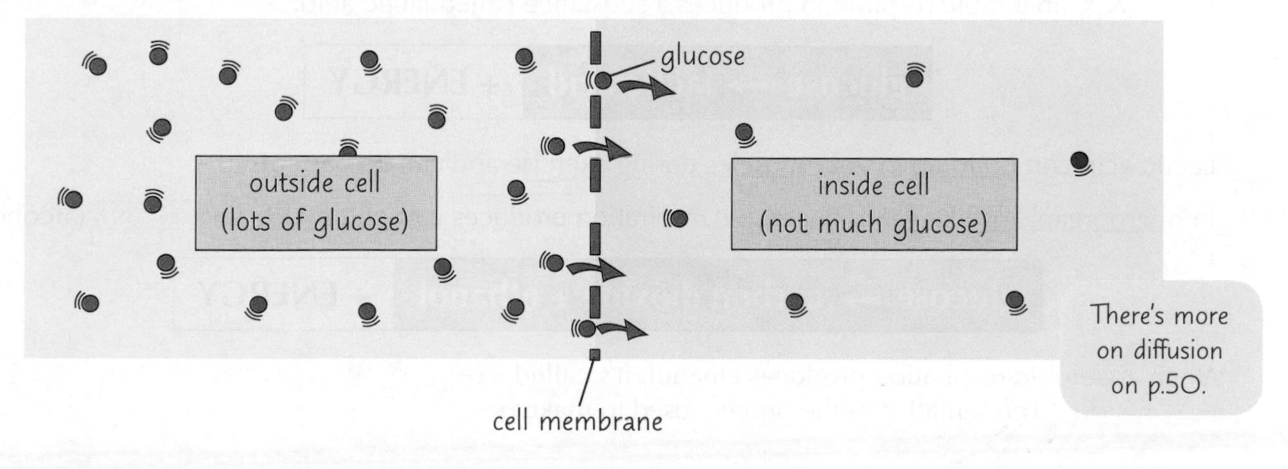

There's more on diffusion on p.50.

Cells are the building blocks of organisms

Remember: cells → tissues → organs → organ systems → organisms. You need to get your head around diffusion too — it comes up all the time in KS3 science, so it's worth getting to grips with now.

Respiration

Respiration is one of the most important life processes there is. It's worth learning really well.

Respiration is a Chemical Reaction

1) Respiration happens in every cell of every living organism.
2) Respiration is the process of releasing energy from glucose (a sugar).
3) The energy released by respiration is used for all the other chemical reactions that keep you alive. For example, the reactions involved in building proteins, muscle contraction and keeping warm.

Aerobic Respiration Needs Plenty of Oxygen

1) Aerobic respiration is respiration using oxygen. It takes place in the mitochondria (see page 2) of animal and plant cells.
2) In aerobic respiration, glucose and oxygen react to produce carbon dioxide and water. This reaction releases lots of energy.

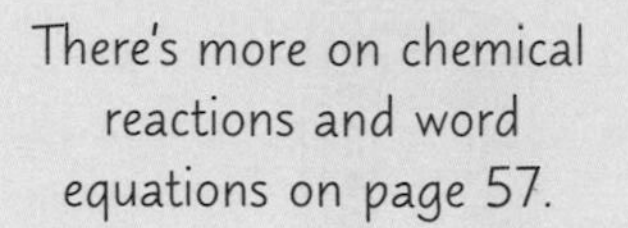

3) Here's a word equation to show what happens in the reaction — learn it:

glucose + oxygen → carbon dioxide + water + ENERGY

These are the reactants. These are the products.

Anaerobic Respiration Takes Place Without Oxygen

1) Anaerobic respiration is respiration without oxygen.
2) Anaerobic respiration is less efficient than aerobic respiration, so it releases less energy.
3) Because of this, anaerobic respiration usually only happens when cells can't get enough oxygen, e.g. if your body can't get enough oxygen to your muscle cells when you exercise, they start to respire anaerobically.

Anaerobic Respiration is Different in Different Organisms

1) In humans, anaerobic respiration produces a substance called lactic acid:

glucose → lactic acid + ENERGY

Lactic acid can build up in your muscles during exercise and can be painful.

2) In microorganisms like yeast, anaerobic respiration produces carbon dioxide and ethanol (alcohol):

glucose → carbon dioxide + ethanol + ENERGY

When anaerobic respiration produces ethanol, it's called fermentation. Fermentation is the process used to make beer.

There are two types of respiration — learn the difference

It can be tricky to get your head around respiration, but it just means turning glucose into energy. Make sure you've learnt those equations — cover the book and write them down.

Warm-Up and Practice Questions

Take a deep breath then ease yourself in gently with these warm-up questions. Then attack the practice questions. All the answers are somewhere in this section, so there are no excuses.

Warm-Up Questions

1) Why would you stain a cell before looking at it under a microscope?
2) Name three structures that are found in both plant and animal cells. Describe what they all do.
3) What is the difference between a tissue and an organ?
4) Which process is responsible for the movement of glucose from an area of high concentration to an area of low concentration?
5) Which sort of respiration involves oxygen? Write the relevant word equation.
6) Which sort of respiration is the most efficient?

Practice Questions

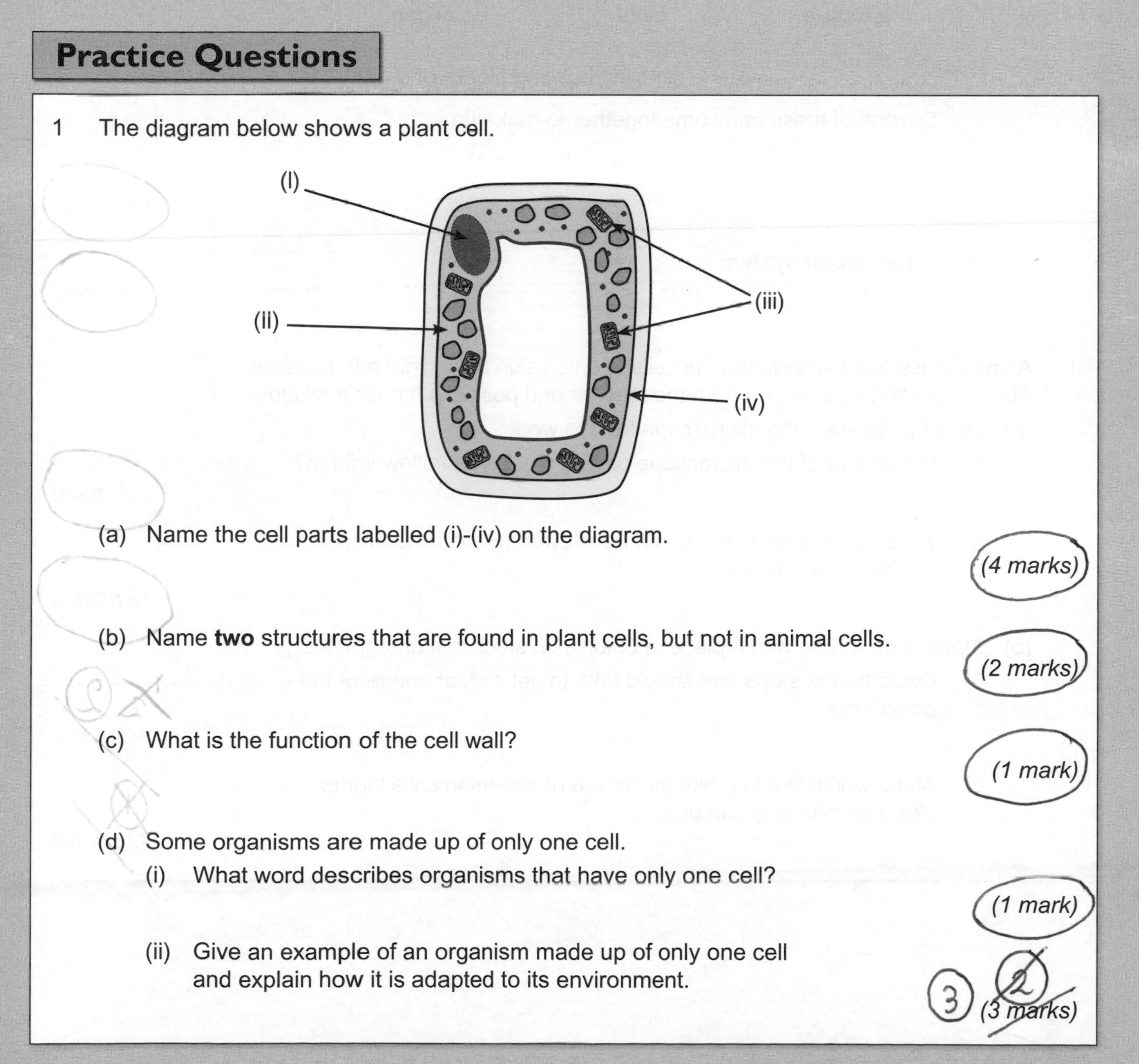

1 The diagram below shows a plant cell.

(a) Name the cell parts labelled (i)-(iv) on the diagram.

(4 marks)

(b) Name **two** structures that are found in plant cells, but not in animal cells.

(2 marks)

(c) What is the function of the cell wall?

(1 mark)

(d) Some organisms are made up of only one cell.

(i) What word describes organisms that have only one cell?

(1 mark)

(ii) Give an example of an organism made up of only one cell and explain how it is adapted to its environment.

(3 marks)

Practice Questions

2 Respiration is a very important life process for all organisms.

(a) In which part of animal and plant cells does aerobic respiration take place?

(1 mark)

(b) Sometimes respiration does not involve oxygen.

(i) Which sort of respiration does not involve oxygen?

(1 mark)

(ii) Write the word equation for this process when it occurs in **humans**.

(1 mark)

(iii) In what situation might a human start respiring in this way?

(1 mark)

3 (a) Use the following words to complete the gaps in the sentences below.

a tissue **cells** **an organ**

.... are the simplest building blocks of organisms.

Several of these can come together to make up .... ,

and several of these can work together to make ....

(3 marks)

(b) What is an **organ system**?

(1 mark)

4 Alana's class are investigating the cells in onion skin using light microscopes.
Alana collects a microscope from the teacher and positions it near a window.

(a) Light has to enter the microscope for it to work.

(i) Which part of the microscope can be adjusted to allow light in?

(1 mark)

(ii) Which kind of light should not be allowed to enter the microscope?
Explain your answer.

(2 marks)

(b) Alana clips a slide with a piece of onion skin stuck to it onto the stage.

(i) Describe the steps she should take to get a clear image of the onion cells.

(4 marks)

(ii) Alana would like to make the image of the onion cells bigger.
Describe how she can do this.

(2 marks)

Revision Summary for Section One

Welcome to your very first Section Summary. It's full of questions written especially for finding out what you actually know — and, more importantly, what you don't. Here's what you have to do...

- Go through the whole lot of these Section Summary questions and try to answer them.
- Look up the answers to any you can't do and try to really learn them (hint: the answers are all somewhere in Section One).
- Try all the questions again to see if you can answer more than you could before.
- Keep going till you get them all right.

1) What part of a microscope do you clip your slide onto?
2) What do the focusing knobs on a microscope do?
3) Why should you always move the objective lens away from the slide when you're focusing a microscope?
4) What is an organism?
5) What instrument would you use to look at a cell?
6) What do chloroplasts do? What sort of cell would you find them in?
7) Explain the meaning of: a) tissue b) organ. Give an example of each.
8) Give an example of an organ system.
9) What is diffusion?
10) Give two examples of substances that move into or out of cells by diffusion.
11) What's the name of the process that goes on in every cell, releasing energy?
12) What is the energy released by this process used for? Give three examples.
13) What is aerobic respiration?
14) Write down all the reactants of aerobic respiration. Now write down the products.
15) Give two differences between aerobic respiration and anaerobic respiration in humans.
16) Write down the word equation for anaerobic respiration in yeast.
17) What is fermentation? What can fermentation be used to make?

Nutrition

Nutrition is what you eat — and what you eat is really important for your health.
A balanced diet will have the right amount of the five nutrients listed below, as well as fibre and water.

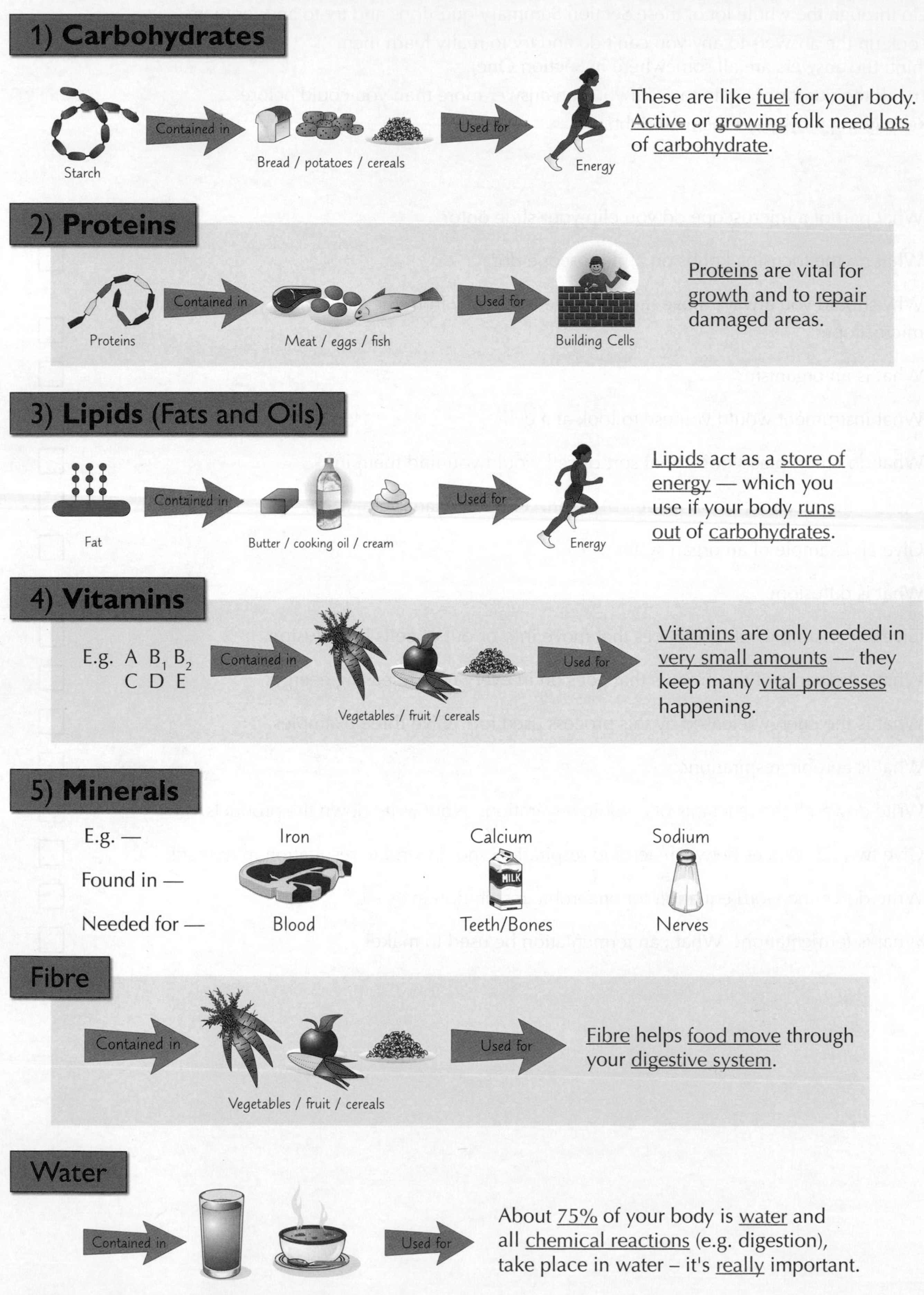

1) Carbohydrates

These are like fuel for your body. Active or growing folk need lots of carbohydrate.

2) Proteins

Proteins are vital for growth and to repair damaged areas.

3) Lipids (Fats and Oils)

Lipids act as a store of energy — which you use if your body runs out of carbohydrates.

4) Vitamins

Vitamins are only needed in very small amounts — they keep many vital processes happening.

5) Minerals

E.g. —	Iron	Calcium	Sodium
Found in —			
Needed for —	Blood	Teeth/Bones	Nerves

Fibre

Fibre helps food move through your digestive system.

Water

About 75% of your body is water and all chemical reactions (e.g. digestion), take place in water – it's really important.

Nutrition and Energy

Your body needs energy all the time. Even when you're asleep your body is using energy just to keep you alive. It's important that you get this energy from a balanced diet, or a few nasty things can happen...

An **Unbalanced Diet** Can Cause **Health Problems**

Obesity

1) If you take in more energy from your diet than you use up, your body will store the extra energy as fat — so you will put on weight.
2) If you weigh over 20% more than the recommended weight for your height, then you are classed as obese.
3) Obesity can lead to health problems such as high blood pressure and heart disease.

Starvation and Deficiency Diseases

1) Some people don't get enough food to eat — this is starvation.
2) The effects of starvation include slow growth (in children), being more likely to get infections, and irregular periods in women.
3) Some people don't get enough vitamins or minerals — this can cause deficiency diseases. For example, a lack of vitamin C can cause scurvy, a deficiency disease that causes problems with the skin, joints and gums.

Different People Have Different **Energy Requirements**

1) The amount of energy you need each day depends on your body mass ("weight") and level of activity.
2) Every cell (see page 2) in the body needs energy. So the bigger you are, the more cells you have, and the more energy you'll need.
3) For every kg of body mass, you need 5.4 kJ of energy every hour. This is the basic energy requirement (BER) needed to maintain essential bodily functions.

A kJ is a unit of energy.

You calculate it like this:

Daily BER (kJ/day) = 5.4 × 24 hours × body mass (kg)

E.g. a 60 kg person requires 5.4 × 24 × 60 = 7776 kJ/day

4) You also need energy to move, and it takes more energy to move a bigger mass.
5) So, the heavier and the more active you are, the more energy you will need.
6) To find out how much energy you need in a day you have to add together your daily BER and the extra energy you use in your activities.

For example, a 60 kg person will use about 400 kJ walking wfor half an hour, but 1500 kJ running for half an hour.

You need to eat a balanced diet to stay healthy

Too much or too little food (or not eating the right foods) can lead to some serious health problems. Make sure you understand the health problems on this page. You also need to know how to work out someone's daily energy requirement — it's important for avoiding the health problems above.

Digestion

Digestion's great. The body breaks down the food we eat, so we can use the nutrients it contains. But it's not easy — lots of different organs have to work together to get the job done.

Digestion is All About Breaking Down Food

There are two steps to this. The first is quick, the second isn't:

1) Breaking down the food MECHANICALLY, e.g. chewing with teeth:

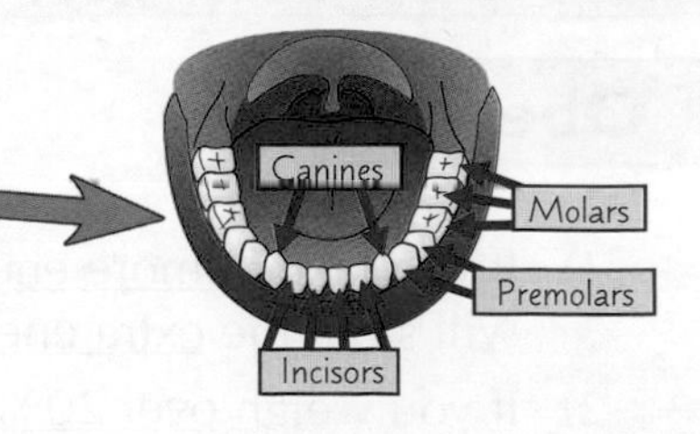

2) Breaking down the food CHEMICALLY — with the help of proteins called enzymes. Enzymes are biological catalysts — this means they speed up the rate of chemical reactions in the body.

Eight Bits of The Alimentary Canal

1) Mouth

Digestion starts here where the teeth have a good old chew and mix the food with saliva. Saliva contains an enzyme (called amylase) that breaks down carbohydrates.

2) Oesophagus

Food pipe — links the mouth to the stomach.

3) Stomach

1) Here the food mixes with protease enzymes which digest proteins. The stomach contains muscular tissue to move the stomach wall and churn up food.
2) Hydrochloric acid is present to kill harmful bacteria and give a low pH for the enzymes to work.

4) Liver

The liver makes bile, which breaks fats into tiny droplets (emulsification). It's also alkaline to give the right pH for the enzymes in the small intestine.

5) Pancreas

The pancreas contains glandular tissue, which makes three enzymes:

1) PROTease digests PROTein.
2) CARBOHYDRAse digests CARBOHYDRAtes.
3) LIPase digests LIPids — i.e. fats.

6) Small intestine

1) This produces more enzymes to further digest proteins, carbohydrates and fats.
2) Food is also absorbed through the gut wall into the blood, which then takes it around the body to wherever it's needed.

7) Large intestine

Here water is absorbed — so we don't all shrivel up.

8) Rectum

Food usually contains some materials that we can't digest. This undigested food is stored as faeces. Here the digestion story ends when it plops out of the anus — egestion.

More on Digestion

Well would you believe it? There's more to learn about digestion.

Absorption of Food Molecules

1) Big, insoluble food molecules can't pass through the gut wall.

2) So enzymes are used to break up the big molecules into smaller, soluble ones.

3) These small molecules can pass through the gut wall into the blood.

4) They are then carried round the body, before passing into cells where they are used.

'Insoluble' means 'won't dissolve'. 'Soluble' means 'will dissolve'. See page 61 for more.

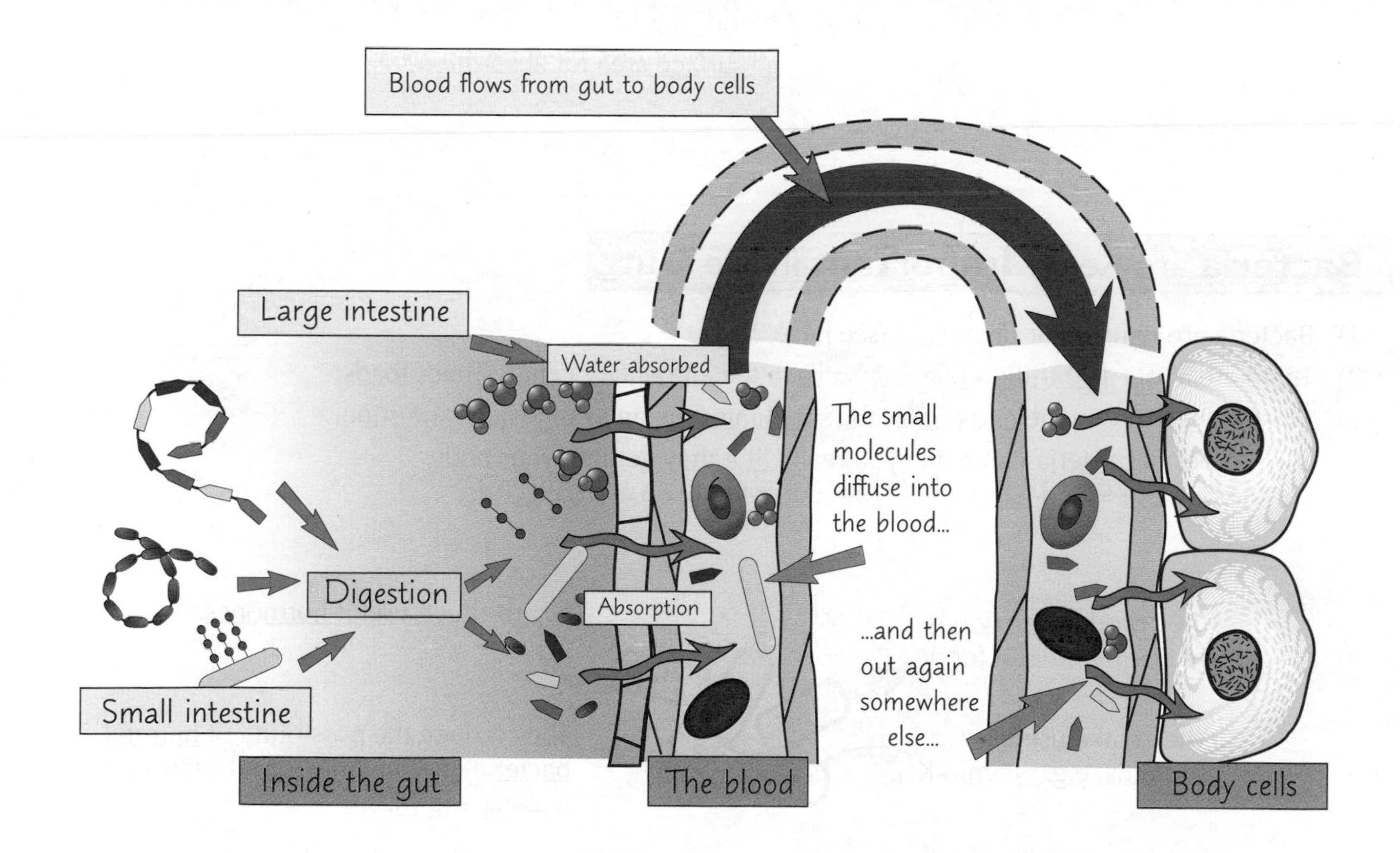

REVISION TASK

You need to absorb all of these facts

As well as looking pretty, the diagrams on digestion are really important for helping you understand how food is broken down and absorbed by the body — so look at them really thoroughly and absorb the information. Make sure you know the name and function of all eight bits of the alimentary canal. To test yourself, cover up these pages and draw the diagrams showing how food is digested. Include as much detail as you can remember.

More on Digestion

More on digestion — don't worry, it's the last page on it, I promise. (Apart from the questions anyway...)

The **Small Intestine** is Covered with **Millions** of **Villi**

1) Food molecules are absorbed into the blood in the small intestine.
2) The small intestine is lined with tiny finger-like projections called VILLI.

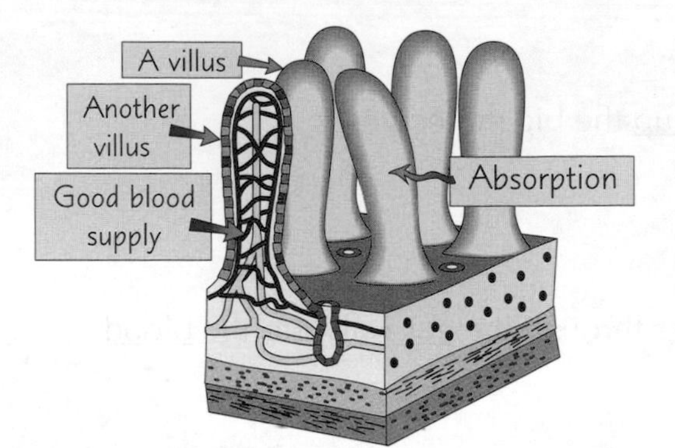

Villi is the plural of villus — i.e. it's one villus but two (or more) villi.

3) Villi are perfect for absorbing food because:

- They have a thin outer layer of cells.
- They have a good blood supply.
- They provide a large surface area for absorption.

Bacteria are Really **Important** in the **Gut**

1) Bacteria are unicellular organisms (see page 2).
2) There are about 100 trillion bacterial cells in the alimentary canal. That's loads.
3) Most of these are in the end part of the small intestine and in the large intestine.
4) Some types of bacteria can make you really ill if they get into your body, but the bacteria found naturally in your gut actually do a lot of good:

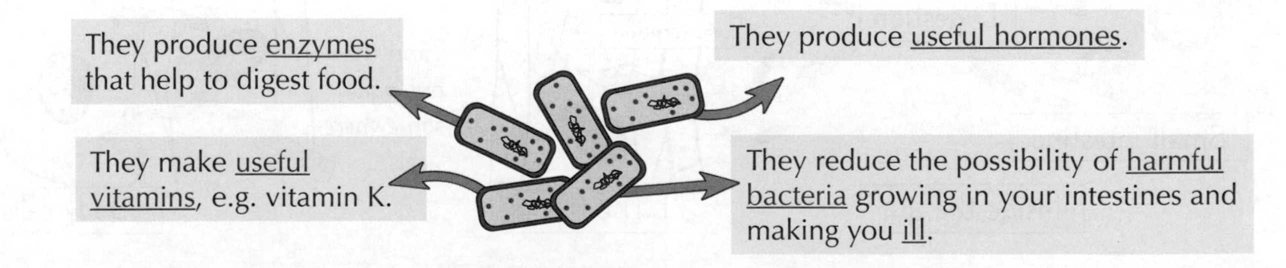

Who knew having bacteria inside you was such a good thing?

Villi are brilliant absorbers of food. Make sure you know the three things that make villi so awesome at doing this — their large surface area, their blood supply and their thin outer layer of cells.

Warm-Up and Practice Questions

When you've digested all that information, have a crack at these questions to test what you know...

Warm-Up Questions

1) Which two nutrients does the body get energy from?
2) Name one type of food that contains fibre.
3) What health risk is caused by taking in more energy than you use up?
4) What is meant by your daily basic energy requirement?
5) Name the two ways that food is broken down by the body.
6) Enzymes are biological catalysts. What does this mean?
7) Which part of the body does digestion start in?
8) Why do we need to digest our food?
9) What type of organisms are present in the gut and produce enzymes that help digest food?

Practice Question

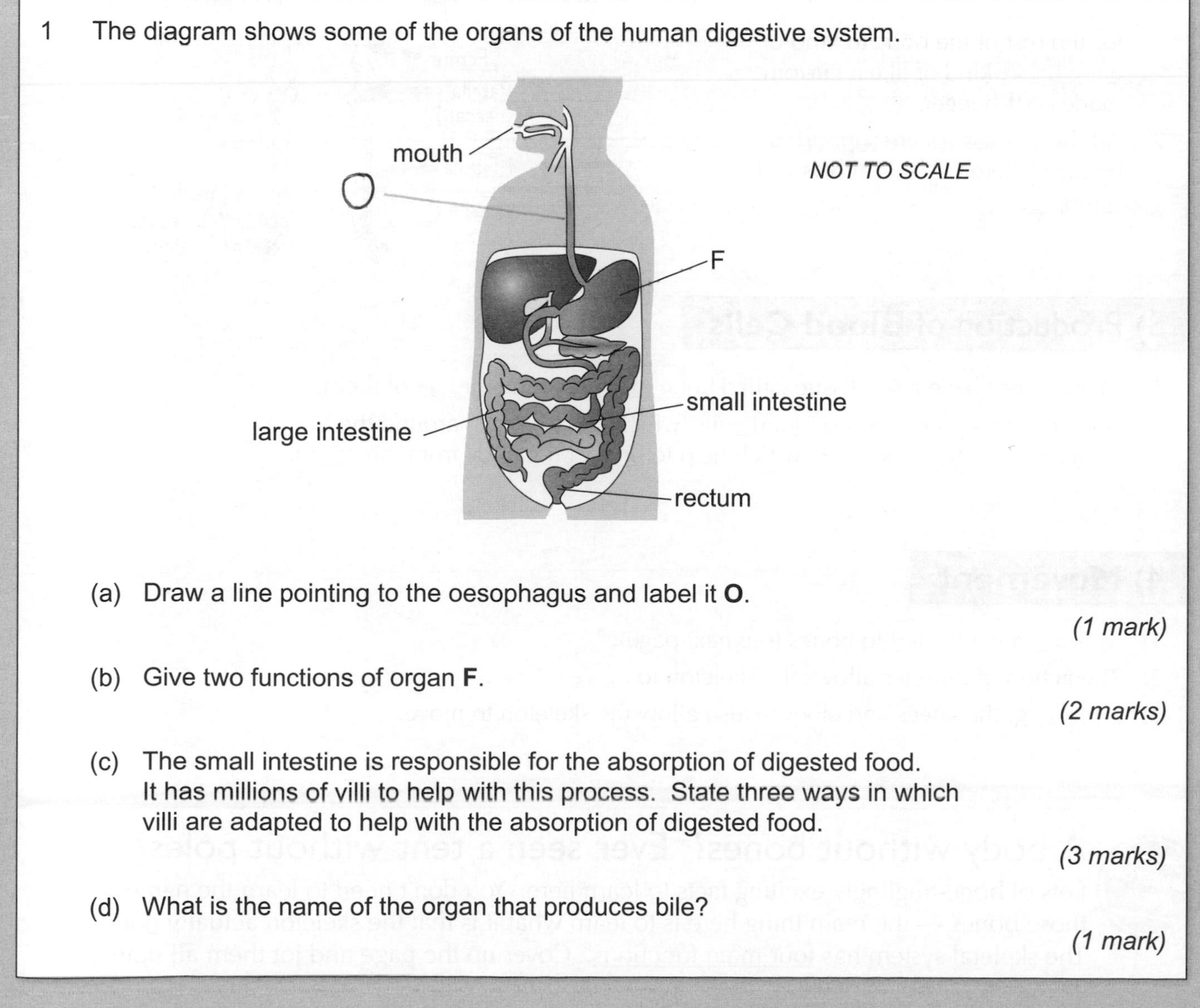

1 The diagram shows some of the organs of the human digestive system.

(a) Draw a line pointing to the oesophagus and label it **O**.

(1 mark)

(b) Give two functions of organ **F**.

(2 marks)

(c) The small intestine is responsible for the absorption of digested food. It has millions of villi to help with this process. State three ways in which villi are adapted to help with the absorption of digested food.

(3 marks)

(d) What is the name of the organ that produces bile?

(1 mark)

The Skeleton

The adult human skeleton is made up of 206 bones. Thankfully you don't need to learn them all...

The Skeletal System

Bones are made from different types of tissue:

- The outer layer of bone is made from really strong and hard tissue — this makes bones rigid (they can't bend).
- The inner layer is made from more spongy tissue, but it's still strong.

The skeletal system has four main functions:

Mandible (Jaw)
Cranium (skull)
Backbone
Clavicle (Collarbone)
Sternum (Breast bone)
Rib
Humerus
Ulna
Radius
Carpals
Metacarpals
Phalanges
Coccyx
Femur
Patella (Kneecap)
Fibula
Tibia
Tarsals
Metatarsals
Phalanges

1) Protection

Bone is rigid and tough so it can protect delicate organs — in particular the brain.

2) Support

1) The skeleton provides a rigid frame for the rest of the body to kind of hang off — kind of like a custom made coat-hanger.
2) All the soft tissues are supported by the skeleton — this allows us to stand up.

3) Production of Blood Cells

1) Many bones have a soft tissue called bone marrow in the middle of them.
2) Bone marrow produces red blood cells (which carry oxygen around the body) and white blood cells (which help to protect the body from infection).

4) Movement

1) Muscles are attached to bones (see next page).
2) The action of muscles allows the skeleton to move.
3) Joints (e.g. the knees and elbows) also allow the skeleton to move.

A body without bones? Ever seen a tent without poles?

Lots of bone-tinglingly exciting facts to learn here. You don't need to learn the names of all those bones — the main thing here is to learn what it is that the skeleton actually does. The skeletal system has four main functions. Cover up the page and jot them all down.

The Muscular System

Another fun number for you — the muscular system is made up of around 640 muscles.
The muscular and skeletal systems work together so you can move around.

Tendons Attach Your Muscles to Your Bones

1) Muscles are attached to bones via tough bands called tendons.
2) When a muscle contracts it applies a force to the bone it's attached to, which makes the bone move.
3) Muscles are found in pairs round a joint (see below).

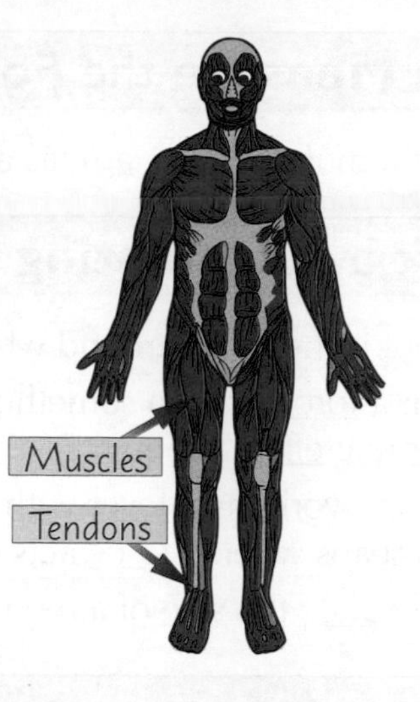

Antagonistic Muscles Work in Pairs

1) Antagonistic muscles are pairs of muscles that work against each other.
2) One muscle contracts (shortens) while the other one relaxes (lengthens) and vice versa.
3) They are attached to bones with tendons. This allows them to pull on the bone, which then acts like a lever (see next page).
4) One muscle pulls the bone in one direction and the other pulls it in the opposite direction — causing movement at the joint.
5) The biceps and triceps muscles in the arm are examples of antagonistic muscles:

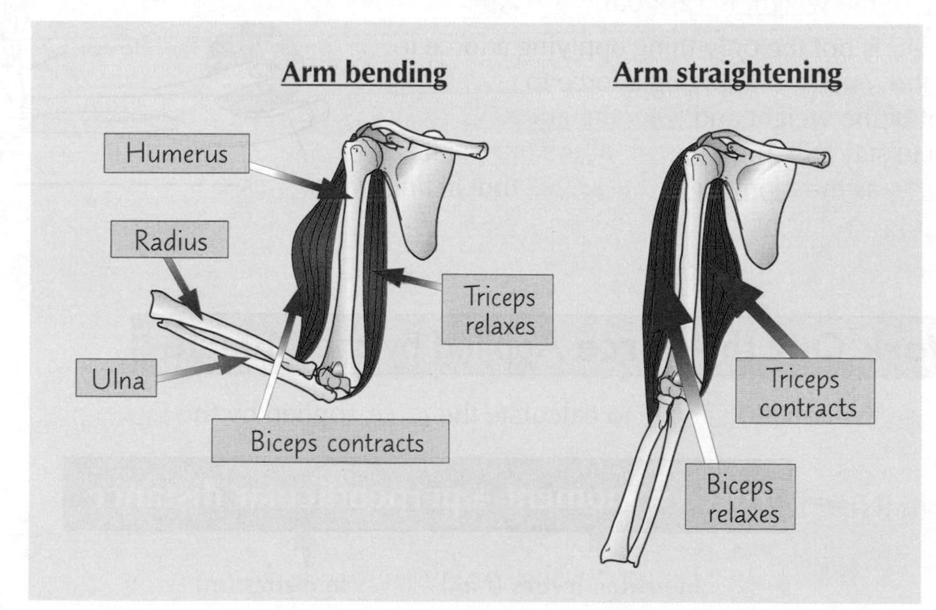

6) The hamstrings and quadriceps in the legs are another example.

When you show off your muscles, you can claim it's revision

Remember that antagonistic muscles just can't get along — whatever one is doing, the other is doing the opposite. Just like me and my sister...

The Force Applied by Muscles

A lot of your bones act like levers that get pulled by muscles. There's a handy little formula you can use to work out how much force a muscle applies to a bone... enjoy.

You Can **Measure** the **Force** Applied by a **Muscle**

The study of forces acting on the body is called biomechanics.

Let's look at a muscle in the arm as an example:

1) Start by **Calculating** the **Moment**

1) A pivot is the point around which a rotation happens. A lever is a bar attached to a pivot.
2) When a force acts on something that has a pivot, it creates a "turning effect" known as a moment (see page 129).
3) The arm works as a lever with the elbow as a pivot. This means when a force acts on the arm there's a moment.
4) To calculate the size of a moment, you can use this equation:

perpendicular distance — force — pivot — a right angle

'Perpendicular distance' is the distance at a right angle from the pivot to the line of force.

Moment = force × perpendicular distance

Moment: In newton metres (Nm); force: In newtons (N); perpendicular distance: In metres (m)

5) In the diagram here, the weight (a force) in the hand is creating a moment.
6) The weight has a force of 12 N. It is 0.3 m away from the pivot (the elbow). So using the equation above, the moment of the weight is 12 × 0.3 = 3.6 Nm.
7) But the weight is not the only thing applying a force to the arm — the muscle is applying a force to counteract the moment of the weight and keep the arm still. For the arm to stay still, the moment of the muscle has to be the same as the moment of the weight (but acting in the opposite direction).

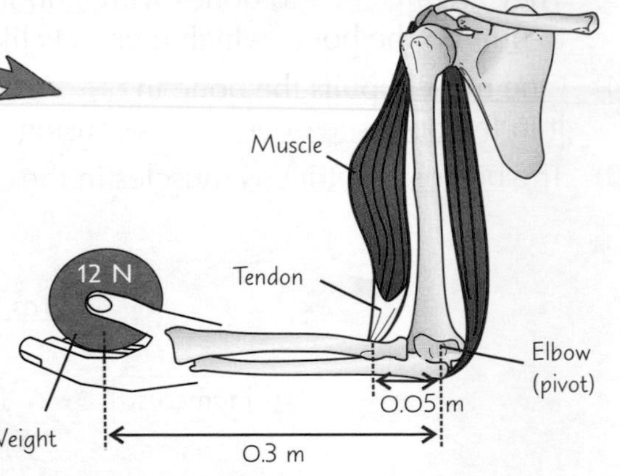

2) Now **Work Out** the **Force** Applied by the **Muscle**

You can rearrange the equation above to calculate the force applied by the muscle:

Force = moment ÷ perpendicular distance

Force: In newtons (N); moment: In newton metres (Nm); perpendicular distance: In metres (m)

In the example above, the weight has a moment of 3.6 Nm, so the muscle must also have a moment of 3.6 Nm.

The distance between the muscle and the pivot (elbow) is 0.05 m. So the force applied by the muscle is 3.6 ÷ 0.05 = 72 N.

Hang on a moment... what?

All this talk of forces and levers and moments can be tricky to get your head around. But stick with it — you'll really impress if you can explain how muscles work and can use that formula to calculate a moment.

Warm-Up and Practice Questions

That's it for bones and muscles. Make sure you know all the things your skeleton does and how your muscles put the whole thing in motion. Time to test how much of the last few pages has made it inside your skull by having a go at these questions...

Warm-Up Questions

1) Name one property of bone that makes it suitable for protecting delicate organs.
2) Describe how the skeleton supports the body.
3) Which part of a bone makes blood cells?
4) What attaches muscles to bones?
5) Describe what antagonistic muscles are and how they work.
6) Give one example of a pair of antagonistic muscles.
7) Write down the equation you would use to calculate the moment caused by a force.

Practice Question

1 The human skeleton has joints with muscles attached to the bones around them, which allow us to move.

(a) Movement is one function of the skeleton.
Write down the other three main functions of the human skeleton.

(3 marks)

(b) The diagram below shows someone holding a box. They hold their arm still.
The weight of the box is 15 N. The distance between the person's elbow joint and the box is 0.35 m. The elbow joint acts as a pivot.

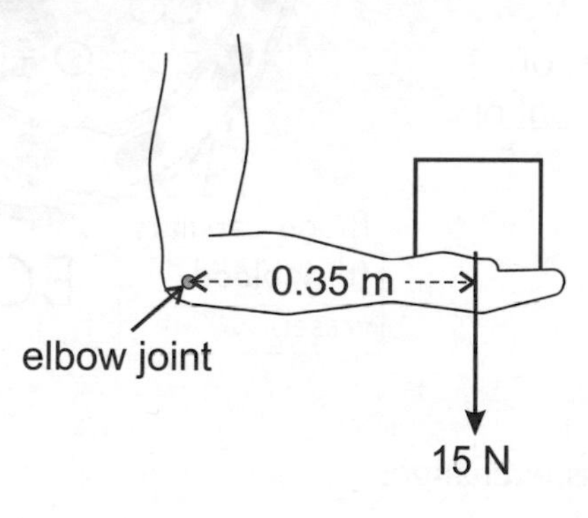

(i) Calculate the moment of the box about the elbow joint.

(1 mark)

(ii) What is the moment of the muscle that is keeping the arm and the box in the position shown? Explain your answer.

(2 marks)

(iii) The distance between the muscle and the elbow joint is 0.05 m. Calculate the force applied by the muscle to keep the arm still in this position.

(1 mark)

Gas Exchange

You need to get oxygen from the air into your bloodstream and to get rid of the carbon dioxide that's in your bloodstream. This all happens in your gas exchange system.

Learn These Structures in the Gas Exchange System

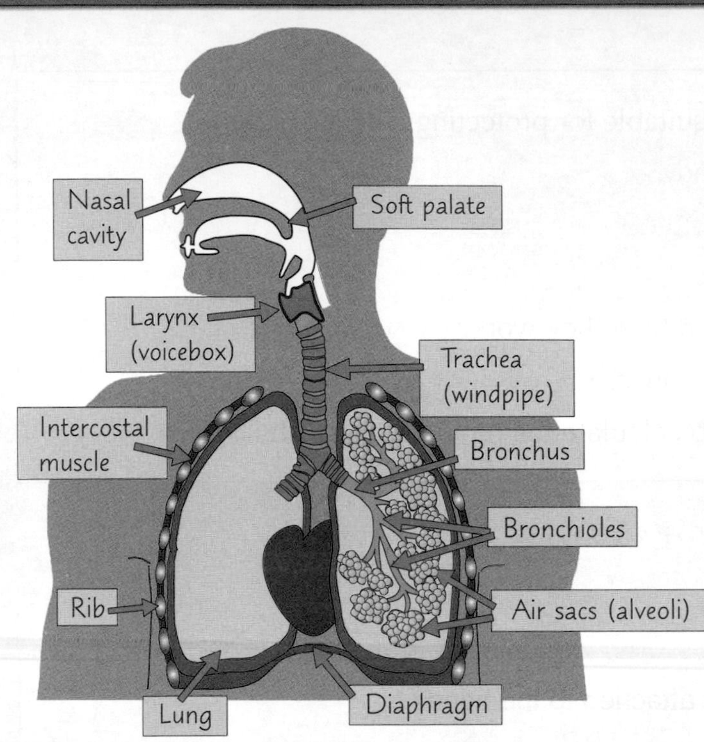

1) The lungs are like big pink sponges. They're protected by the ribcage.
2) The diaphragm is a muscle that sits underneath the ribcage. It moves up when it relaxes and down when it contracts. This movement helps to get air in and out of your lungs (see next page).
3) The air that you breathe in goes through the trachea. This splits into two tubes called 'bronchi' (each one is 'a bronchus'), one going to each lung.
4) The bronchi split into smaller tubes called bronchioles.
5) The bronchioles end at small air sacs in the lungs called alveoli. These are where gas exchange takes place.

Gas Exchange Happens in the Lungs

1) Air is inhaled into the lungs.
2) Some of the oxygen in the inhaled air passes into the bloodstream to be used in respiration (see page 4).
3) Carbon dioxide is a waste product of respiration. In the lungs it passes out of the blood and is then breathed out.
4) The gases pass into or out of the bloodstream by diffusion — where a substance moves from where there's lots of it to where there's less of it (see page 3).
5) The lungs are well adapted for gas exchange:

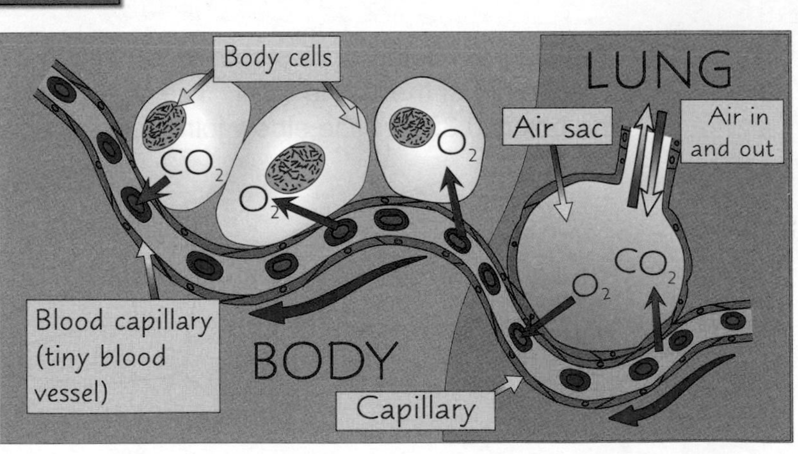

1) They're moist.
2) They have a good blood supply.
3) The alveoli (air sacs) give the lungs a big inside surface area.

I love ribs — spare ones are my favourite though

There are a couple of detailed diagrams here which need learning. Sooner or later you're expected to learn all the structures in the gas exchange system and what they do, so you may as well start now.

Breathing

Breathing is how the air gets in and out of your lungs. It's definitely a useful skill.

The Mechanism of Breathing

Air rushes in
Balloons fill up like lungs
Pull down

The bell jar demonstration shows us what's going on when you breathe:

1) First you pull the rubber sheet down — like it's your diaphragm.
2) This increases the volume inside the bell jar, which decreases the pressure.
3) The drop in pressure causes air to rush into the balloons — this is like breathing in.
4) Let go of the rubber sheet — this is like relaxing your diaphragm.
5) The volume in the jar gets smaller. This increases the pressure, so air rushes out.

Air rushes out
Balloons deflate
Relax back up

Inhaling and Exhaling is Breathing In and Out

1) The chest cavity is like a bell jar.
2) When you breathe in, the diaphragm moves down and the ribs move up. This increases the volume of the chest cavity, which decreases the pressure. So air rushes in to fill the lungs.
3) When the diaphragm moves up and the ribs move down, air rushes out.

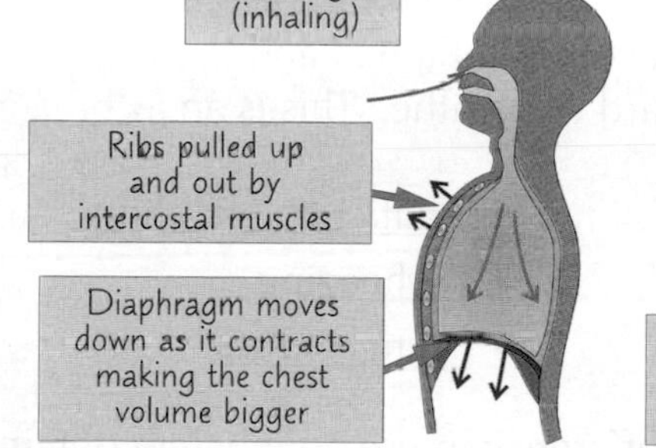

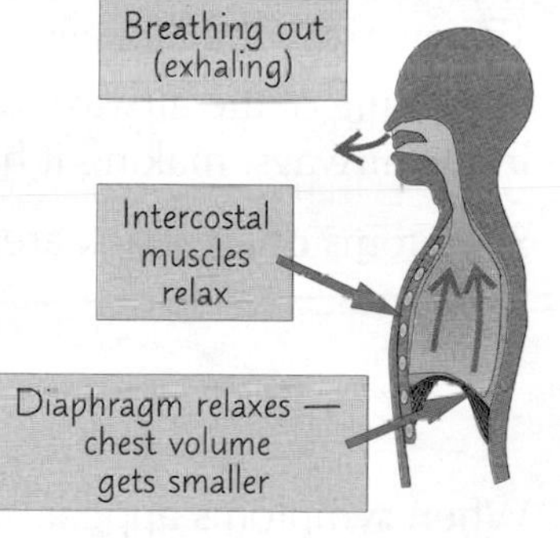

Lung Volume Can Be Measured

1) Lung volume is the amount of air you can breathe into your lungs in a single breath.
2) Lung volume is different for different people. For example, taller people tend to have a bigger lung volume than shorter people. And some diseases may reduce a person's lung volume.
3) Lung volume can be measured using a machine called a spirometer.
4) To use a spirometer, a person breathes into the machine (through a tube) for a few minutes.
5) The volume of air that is breathed in and out is measured and plotted on a graph (called a spirogram) like this one.

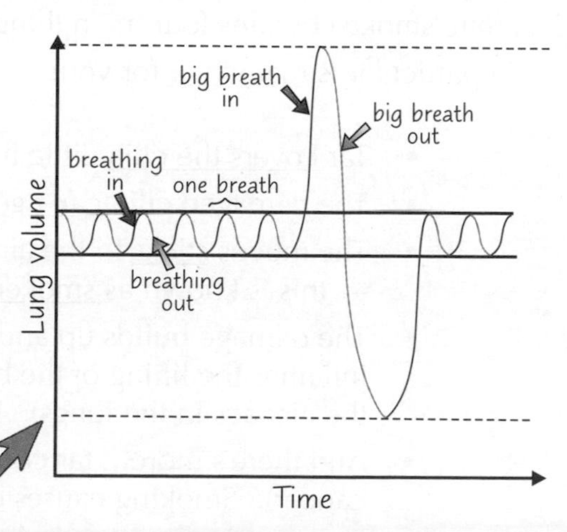

Now take a deep breath and learn these facts

Well, if ever you wanted to know how you breathe in and out, now you do. Learn how breathing works — use that bell jar demonstration to help you understand what goes on in your actual lungs. Make sure you know how lung volume can be measured too. You'll be an expert in breathing soon.

Exercise, Asthma and Smoking

Exercise, asthma and smoking can all affect your gas exchange system and the way that you breathe.

Exercise

1) When you exercise, your muscles need more oxygen and glucose so they can respire and release energy (see page 4) to keep you going.
2) During exercise, your breathing rate and depth of breathing increase so you can get more oxygen into your blood.
3) If you exercise regularly, the muscles that you use to breathe (the diaphragm and intercostal muscles) will get stronger.
4) This means that your chest cavity can open up more when you breathe in, so you can get more air into your lungs.
5) Over time, regular exercise can also cause an increase in the number and size of the small blood vessels in your lungs and in the number of alveoli. This makes gas exchange more efficient.

Asthma

1) People with asthma (asthmatics) have lungs that are too sensitive to certain things (e.g. pet hair, pollen, dust, smoke...).
2) If an asthmatic breathes these things in, the muscles around their bronchioles contract. This narrows the airways.
3) The lining of the airways becomes inflamed and fluid builds up in the airways, making it hard to breathe. This is an asthma attack.
4) Symptoms of an attack are:
   - difficulty breathing,
   - wheezing,
   - a tight chest.
5) When symptoms appear, sufferers can use an inhaler containing drugs that open up the airways.

Smoking

1) Cigarette smoke contains four main things: carbon monoxide, nicotine, tar and particulates.
2) Tar in particular is really bad for you:
   - Tar covers the cilia (little hairs) on the lining of the airways.
   - The damaged cilia can't get rid of mucus properly.
   - The mucus sticks to the airways, making you cough more — this is known as smoker's cough.
   - The damage builds up and can eventually lead to bronchitis (a disease that inflames the lining of the bronchi) and emphysema (a disease that destroys the air sacs in the lungs). Both these diseases make it difficult to breathe.
   - And there's more... tar contains carcinogens (substances that can cause cancer). Smoking causes cancer of the lung, throat and mouth.

This page is just breathtaking

So there you have it, three different things that have an impact on the gas exchange system. Make sure you get to grips with all of them — cover up the page and see how much you can write about each one.

Warm-Up and Practice Questions

Take a deep breath, then have a bash at these questions...

Warm-Up Questions

1) Why is it important to have a good blood supply going to the lungs?
2) What is meant by lung volume?
3) What does a spirometer measure?
4) Give two changes that can happen to your gas exchange system if you exercise regularly over a long period of time.
5) Explain why a person's airways narrow during an asthma attack.

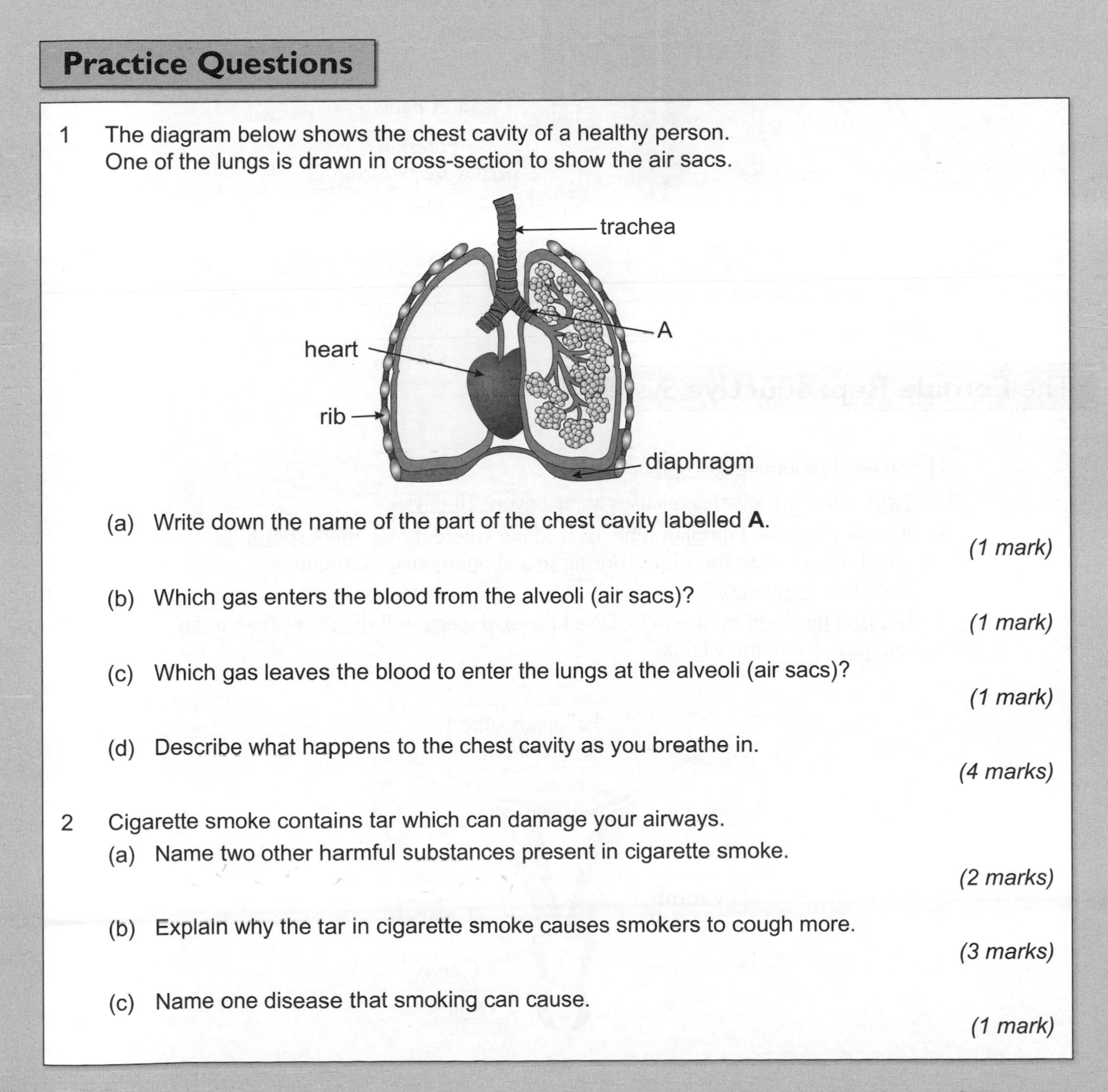

Practice Questions

1 The diagram below shows the chest cavity of a healthy person.
One of the lungs is drawn in cross-section to show the air sacs.

(a) Write down the name of the part of the chest cavity labelled **A**.

(1 mark)

(b) Which gas enters the blood from the alveoli (air sacs)?

(1 mark)

(c) Which gas leaves the blood to enter the lungs at the alveoli (air sacs)?

(1 mark)

(d) Describe what happens to the chest cavity as you breathe in.

(4 marks)

2 Cigarette smoke contains tar which can damage your airways.

(a) Name two other harmful substances present in cigarette smoke.

(2 marks)

(b) Explain why the tar in cigarette smoke causes smokers to cough more.

(3 marks)

(c) Name one disease that smoking can cause.

(1 mark)

Human Reproductive System

Like all mammals, we have different male parts and female parts that allow us to reproduce. No giggling...

The Male Reproductive System

1) Sperm are the male sex cells or 'gametes'.
2) Sperm are made in the testes after puberty.
3) Sperm mix with a liquid to make semen, which is ejaculated from the penis during sexual intercourse.

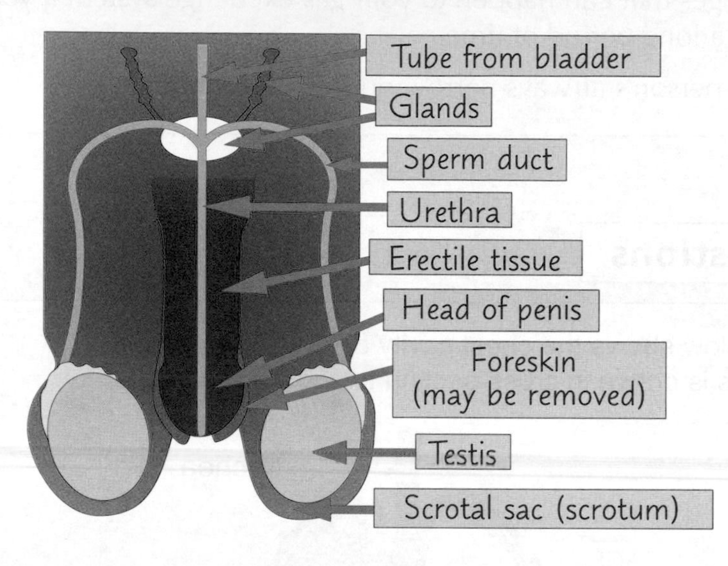

The Female Reproductive System

1) An egg is a female sex cell or 'gamete'.
2) One of the two ovaries releases an egg every 28 days.
3) It passes into the fallopian tube (or oviduct) where it may meet sperm, which has entered the vagina during sexual intercourse (sometimes known as copulation).
4) If it isn't fertilised by sperm (see next page), the egg will die after about a day and pass out of the vagina.

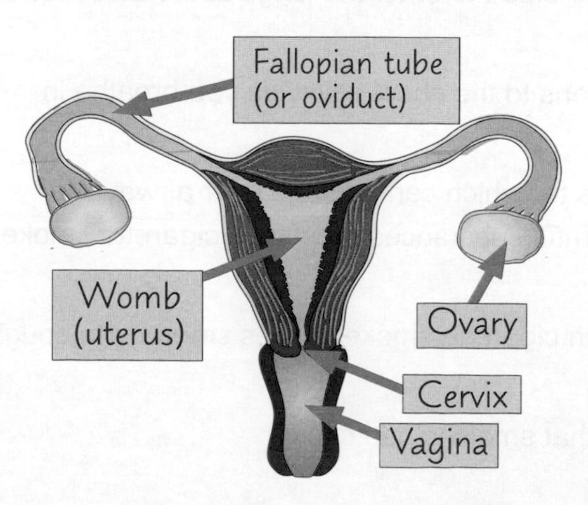

The Menstrual Cycle

The menstrual cycle — not the most exciting of things, but you wouldn't be here without it.

The Menstrual Cycle Takes 28 Days

1) From the age of puberty, females undergo a monthly sequence of events which are collectively known as the MENSTRUAL CYCLE.

2) This involves the body preparing the uterus (womb) in case it receives a fertilised egg.

3) If this doesn't happen, then the egg and uterus lining break down and are lost from the body through the vagina over a period of three to four days, usually.

4) The cycle has four main stages — they are summarised in the diagram and table below:

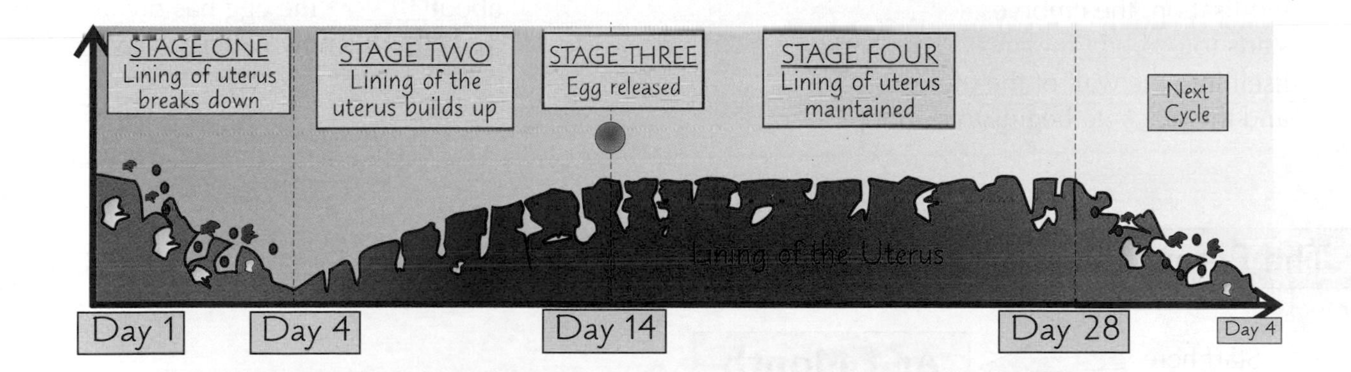

Day	What happens...
1	BLEEDING STARTS as the lining of the uterus (the womb) breaks down and passes out of the vagina — this is what's known as "having a PERIOD".
4	The lining of the uterus starts to build up again. It thickens into a spongy layer full of blood vessels ready for IMPLANTATION. (See next page.)
14	An egg is released from the ovaries of the female, so this is the MOST LIKELY time in which a female may become pregnant. (This day may vary from one woman to the next.)
28	The wall remains thick awaiting the arrival of a fertilised egg. If this doesn't happen then this lining breaks down, passing out of the vagina. Then the whole cycle starts again.

Menstruation — nothing to do with 'men' whatsoever

Phew, there are quite a few details to learn here. Make sure you know the names of all the bits and bobs in the male and female reproductive systems on page 22. You need to know exactly what happens at each of the four stages of the menstrual cycle and when they occur too.

Having a Baby

Once Dad's sperm has fertilised Mum's egg, an embryo forms, gestation happens, and a baby is born.

Fertilisation and Development

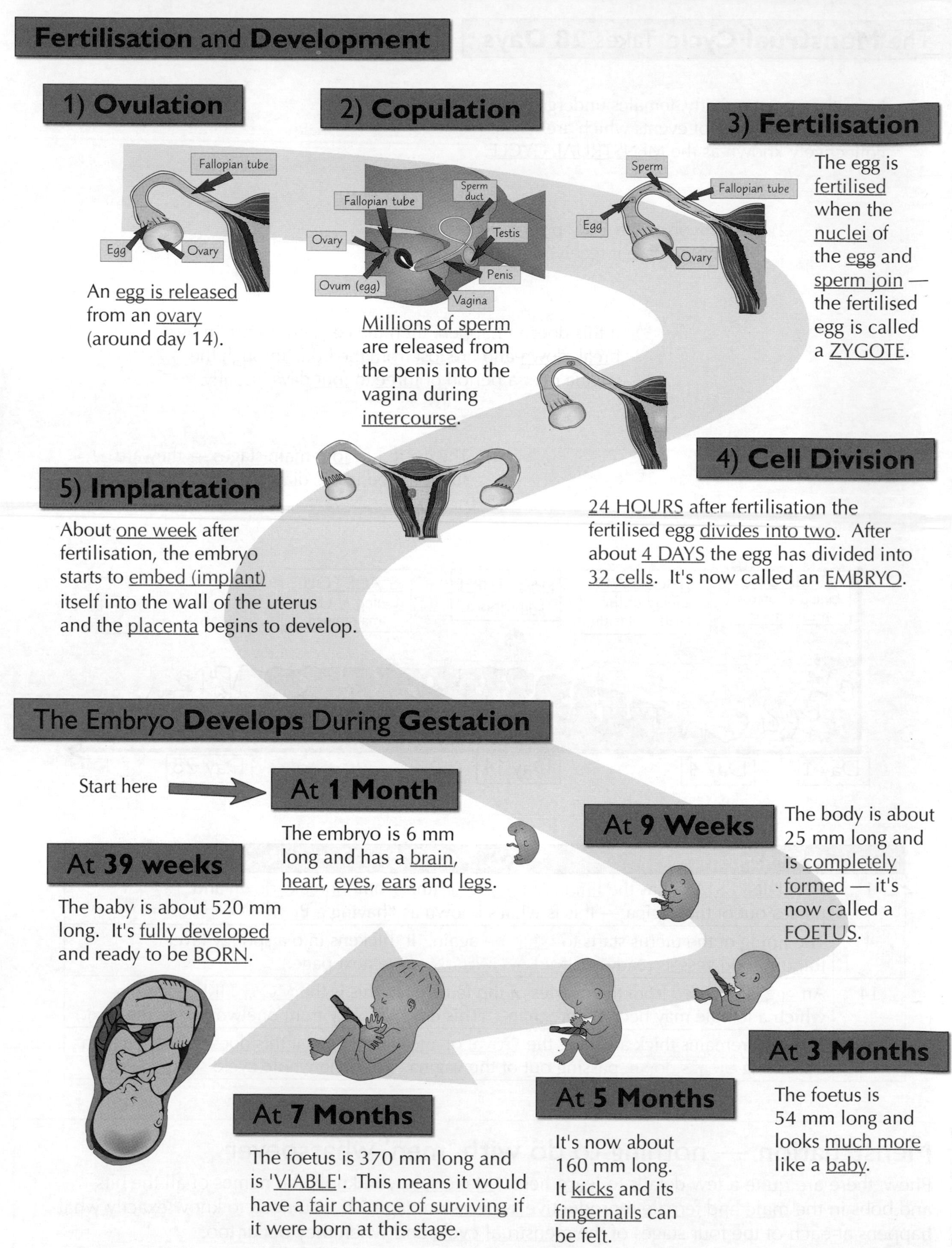

1) Ovulation

An egg is released from an ovary (around day 14).

2) Copulation

Millions of sperm are released from the penis into the vagina during intercourse.

3) Fertilisation

The egg is fertilised when the nuclei of the egg and sperm join — the fertilised egg is called a ZYGOTE.

4) Cell Division

24 HOURS after fertilisation the fertilised egg divides into two. After about 4 DAYS the egg has divided into 32 cells. It's now called an EMBRYO.

5) Implantation

About one week after fertilisation, the embryo starts to embed (implant) itself into the wall of the uterus and the placenta begins to develop.

The Embryo Develops During Gestation

At 1 Month

The embryo is 6 mm long and has a brain, heart, eyes, ears and legs.

At 9 Weeks

The body is about 25 mm long and is completely formed — it's now called a FOETUS.

At 3 Months

The foetus is 54 mm long and looks much more like a baby.

At 5 Months

It's now about 160 mm long. It kicks and its fingernails can be felt.

At 7 Months

The foetus is 370 mm long and is 'VIABLE'. This means it would have a fair chance of surviving if it were born at this stage.

At 39 weeks

The baby is about 520 mm long. It's fully developed and ready to be BORN.

Health and Pregnancy

Good health is a situation where you're fine and dandy both physically and mentally. It's important to make sure you look after your health if you're pregnant, as your health affects the baby's health.

Health is More Than Just the Absence of Disease

Good health means having BOTH of these:

1) A healthy body that's all working properly with no diseases.
2) A healthy mental state where you're able to cope with the ups and downs of life.

You should look after your body by eating a balanced diet, doing enough exercise and not abusing drugs.

The Mother's Lifestyle During Pregnancy is Important

1) The placenta lets the blood of the foetus and mother get very close to allow exchange of food, oxygen and wastes.

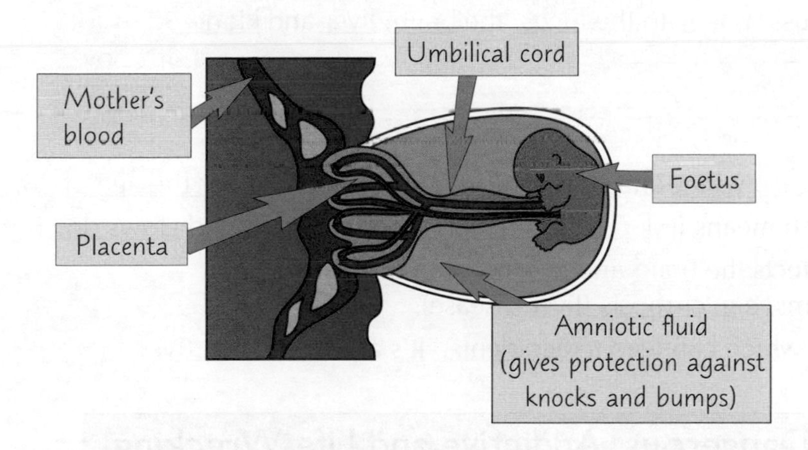

2) If the mother smokes, drinks alcohol or takes other drugs while she is pregnant, harmful chemicals in her blood can cross the placenta and affect the foetus.

3) For example, the foetus may not develop properly and could have health problems after it's born.

Well, that's all a bit different to the stork story I got told

There's a lot to learn on these pages. Make sure you know all the different stages it takes to make an embryo and how an embryo develops into a baby ready to pop out into the world. Remember, having good health and a good lifestyle is important for everyone, but especially when you're pregnant.

Drugs

Recreational drugs can have serious negative effects on your health.

Drugs

1) A drug is any substance that affects the way the body works. E.g. They may raise the heart rate or cause blurred vision.
2) There are LEGAL DRUGS and ILLEGAL DRUGS. Aspirin, caffeine and antibiotics are examples of legal drugs. Cannabis, speed and ecstasy are examples of illegal drugs.
3) RECREATIONAL DRUGS are drugs used for fun. They can be legal or illegal.
4) Drugs can affect life processes. For example, drugs that affect the brain are likely to affect movement and sensitivity. And drugs that affect the liver and kidneys will most likely affect excretion (as these are the organs that process waste).

The 7 Life Processes

Movement — moving parts of the body.
Reproduction — producing offspring.
Sensitivity — responding and reacting.
Nutrition — getting food to stay alive.
Excretion — getting rid of waste.
Respiration — turning food into energy.
Growth — getting to adult size.

Solvents

1) Solvents are found in most homes — in things like paints, aerosols and glues.
2) They're drugs because they cause hallucinations, which are illusions of the mind. Solvents usually have a severe effect on behaviour and character.
3) They also cause serious damage to the lungs, the brain, liver and kidneys.

Alcohol

1) Alcohol is found in beers, wines and spirits. It's illegal to buy it under the age of 18.
2) It's a depressant, which means it decreases the activity of the brain and slows down responses.
3) It's a poison which affects the brain and liver leading to various health problems, e.g. cirrhosis (liver disease).
4) It impairs judgement, which can lead to accidents. It's also very addictive.

Illegal Drugs — Dangerous, Addictive and Life-Wrecking

1) Ecstasy and LSD are hallucinogens. Ecstasy can give the feeling of boundless energy which can lead to overheating, dehydration and sometimes DEATH.
2) Heroin and morphine were developed as painkillers. However they turned out to be highly addictive. They can both cause severe degeneration of a person's life.
3) Amphetamine (speed) and methedrine are stimulants. They give a feeling of boundless energy. However, users quickly become psychologically dependent on the drug (i.e. they think they need them), so behaviour and character deteriorate.
4) Barbiturates are depressants. They slow down the nervous system and therefore slow down reaction time. They can help you to sleep but they're seriously habit-forming.

Drugs aren't harmless fun — they're a slippery slope

It's important that you know the different effects that drugs can have on your health. Make sure that you know how different types of recreational drugs can affect behaviour, health and life processes — use MRS NERG to remember the 7 life processes.

Warm-Up and Practice Questions

Well, that's almost it for this section. Just a few questions to go and you're done.

Warm-Up Questions

1) What are the male sex cells called? Where are they made?
2) On which day, approximately, will an egg be released from the ovary, during a 'normal' 28-day menstrual cycle?
3) State one function of the placenta, as the embryo develops inside the mother's uterus.
4) You need a healthy body to have good health. What else do you need?
5) Explain why it's not a good idea for a woman to smoke while she's pregnant.
6) What is meant by a 'recreational' drug?
7) Name two things containing solvents that you can find in the home.
8) Name two organs that you can damage by using solvents.
9) Name two drugs that are hallucinogens.

Practice Questions

1 The diagram below shows the human female reproductive system.

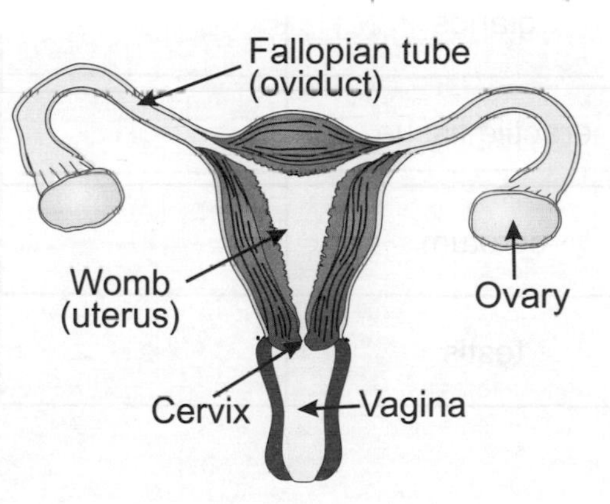

(a) A female will usually release an egg from an ovary roughly every 28 days. What is this process called?

(1 mark)

(b) (i) In what part of the female reproductive system does fertilisation usually take place?

(1 mark)

(ii) Underline the correct definition of **fertilisation** in the list below:

When an egg cell is released from the ovary

When the egg and sperm meet

When the nuclei of the egg and sperm join

When the egg and sperm attach to the uterus wall

(1 mark)

(c) After how many weeks of pregnancy is a human baby considered to be 'fully developed'?

(1 mark)

Practice Questions

2 (a) Five parts of the human male reproductive system are named in the table below. Using the diagram, write the letter for each part next to its name.

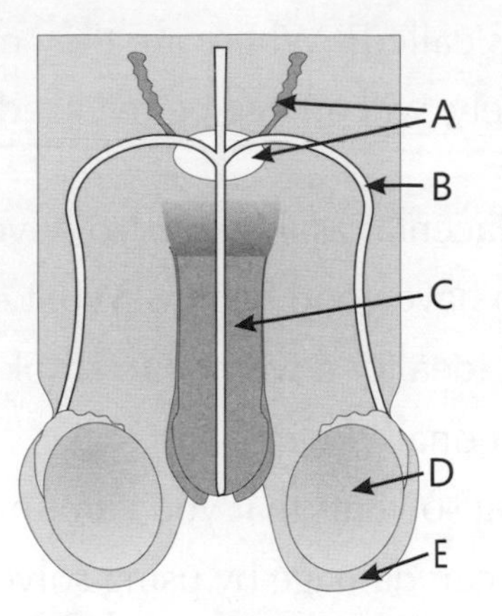

name of organ	letter
sperm duct	
glands	
erectile tissue	
scrotum	
testis	

(5 marks)

(b) What is the name of the substance ejaculated from the penis during sexual intercourse?

(1 mark)

3 Alcohol is one type of legal drug.

(a) Name one other type of legal drug.

(1 mark)

(b) Why does drinking alcohol slow down a person's reactions?

(1 mark)

(c) Write down **two** organs in the body that can be damaged by drinking alcohol.

(2 marks)

(d) Alcohol is a depressant. Name **one** other type of drug that is a depressant.

(1 mark)

Revision Summary for Section Two

Well, that's the end of Section Two. Now what you've got to do is make sure you learn it all. And here again for your enjoyment are some more of those splendid questions. Remember, you have to keep coming back to these questions time and time again, to see how many of them you can do. All they do is test the basic simple facts. OK then — let's see how much you've learnt so far...

1) Name all five nutrients in a balanced diet. Say what each nutrient is important for in the body. ☐
2) For each of the five nutrients, give three examples of foods that contain them. ☐
3) Apart from the five nutrients, give two things that are needed in a balanced diet and explain why they're needed. ☐
4) What is obesity? How is it caused? ☐
5) What health problems can be caused by getting too little food? ☐
6) Give two things that affect how much energy a person needs each day. ☐
7)* Sonia has a body mass of 54 kg. What is her daily basic energy requirement? ☐
8) Name eight main bits of the alimentary canal. Say what goes on in each of the eight bits. ☐
9) Why can't big molecules pass through gut walls? What has to happen to them first? ☐
10) What are villi? What is their function (job) and how are they well-suited to do it? ☐
11) Give four reasons why the bacteria found naturally in your digestive system are good news. ☐
12) Explain how the skeleton protects parts of the body. ☐
13) What are antagonistic muscles? ☐
14) Explain in terms of "muscle contraction" how you can move your arm up and down. ☐
15) What is a moment? What two pieces of information do you need to be able to calculate one? ☐
16) Sketch the human gas exchange system and label all the important structures. ☐
17) What gases are exchanged in the lungs when air is breathed in? Where does each gas move from and to? ☐
18) Give three ways in which the lungs are well-adapted for gas exchange. ☐
19) Explain how we breathe air in and out. ☐
20) How can lung volume be measured? ☐
21) What happens in the gas exchange system when someone has an asthma attack? ☐
22) What are the symptoms of an asthma attack? ☐
23) Give two ways in which smoking affects the gas exchange system. ☐
24) What are the human female sex cells called? Where are they made? ☐
25) Outline the four main stages of the menstrual cycle and say when they happen. ☐
26) In human reproduction, what is meant by implantation? ☐
27) Describe what an embryo looks like at: 1 month, 9 weeks, 3 months, 5 months, 7 months, 39 weeks. ☐
28) What does being 'healthy' mean? ☐
29) What are drugs? Name three legal drugs and three illegal drugs. ☐
30) Name one recreational drug and explain how it affects life processes. ☐

*Answer on page 189.

Plant Nutrition

Think about this: plants make their own food — it's a nice trick if you can do it.

Photosynthesis Makes Food From Sunlight

1) Photosynthesis is a chemical process which takes place in every green plant.
2) Photosynthesis basically produces food — in the form of glucose (a carbohydrate).
3) The plant can then use the glucose to increase its biomass — i.e. to grow.
4) Photosynthesis happens in all the green bits of a plant but mainly in the leaves.

Four Things are Needed for Photosynthesis...

1) Sunlight
2) Chlorophyll — a green chemical found in the chloroplasts of plant cells.
3) Carbon dioxide — this diffuses into the leaves from the air.
4) Water — this is absorbed from the soil by the plant roots and is carried up to the leaves.

There's more on chloroplasts on p.2.

The chlorophyll absorbs sunlight and uses its energy to convert carbon dioxide and water into glucose. Oxygen is also produced. This word equation summarises what happens during photosynthesis. Learn it:

Carbon dioxide + Water —(Sunlight, Chlorophyll)→ Glucose + Oxygen

These are the reactants. These are the products.

Leaves are Adapted for Efficient Photosynthesis

Leaves are really good at carrying out photosynthesis. Here's why...

1) Leaves are broad, so there's a big surface area for absorbing light.
2) Most of the chloroplasts are found in cells near the top of the leaf, where they can get the most light.
3) The underside of the leaf is covered in tiny holes called stomata. These holes allow carbon dioxide to diffuse (move) into the leaf from the air. They also allow oxygen to diffuse out. Air spaces inside the leaf allow carbon dioxide to move easily between the leaf cells.
4) Leaves also contain a network of veins, which deliver water to the leaf cells and take away glucose.

Plants Absorb Minerals from the Soil Through Their Roots

1) Plants grow using the food they make themselves in photosynthesis. But to keep healthy they also need mineral nutrients from the soil.
2) Plants absorb these minerals through their roots (along with water).

Hmm, it's all clever stuff — just make sure you learn it

Remember, plants make their own food using photosynthesis. Chlorophyll absorbs sunlight and uses the energy to make glucose and oxygen from carbon dioxide and water. The roots suck up all the water needed for photosynthesis as well as nutrients needed from the soil. Got that? Sorted.

Plant Reproduction

The **Flower** Contains the **Reproductive Organs**

1) Stamens

The sta-men-s are the male parts of the flower. They consist of the anther and the filament. The anther contains pollen grains, which produce the male sex cells. The filament supports the anther.

2) Carpels

The female parts of the flower. They consist of the stigma, style and ovary. The ovary contains the female sex cells inside ovules.

3) Petals

These are often brightly coloured. They attract the insects needed for pollination.

4) Sepals

These are green and leaf-like. They protect the flower in the bud. They're found below the main petals.

"**Pollination**" Is Getting **Pollen** to the **Stigma**

1) To make a seed (which will eventually grow into a new plant) the male and female sex cells must "meet up".
2) To do this, the pollen grains must get from a stamen to a stigma. This can happen in two ways:

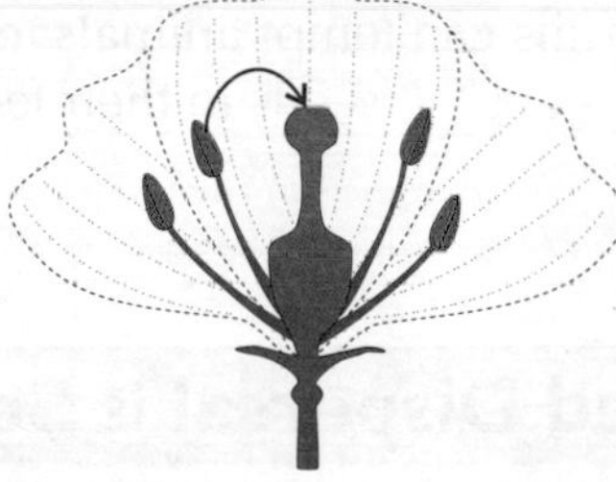

1) Self-pollination — pollen is transferred from stamen to stigma on the SAME PLANT.

2) Cross-pollination — pollen is transferred from the stamen of one plant to the stigma of a DIFFERENT PLANT. Cross-pollination can involve...

...**Insect** Pollination

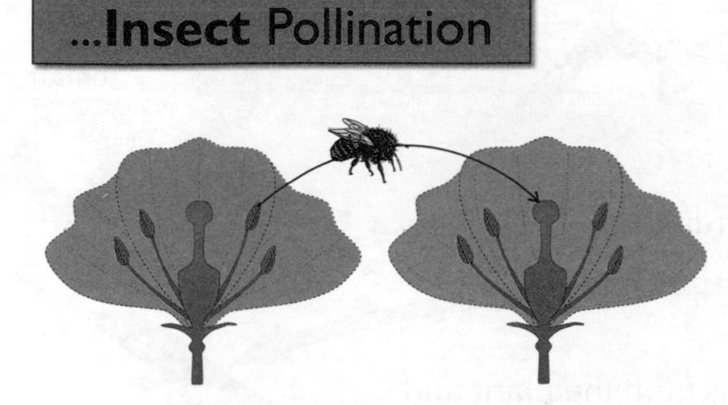

Plant features that help insect pollination:

1) Bright coloured petals.
2) Scented flowers with nectaries (glands that produce a sugary liquid for insects to feed on).
3) Sticky stigma to take the pollen off the insect as it goes from plant to plant to feed in the nectaries.

...**Wind** Pollination

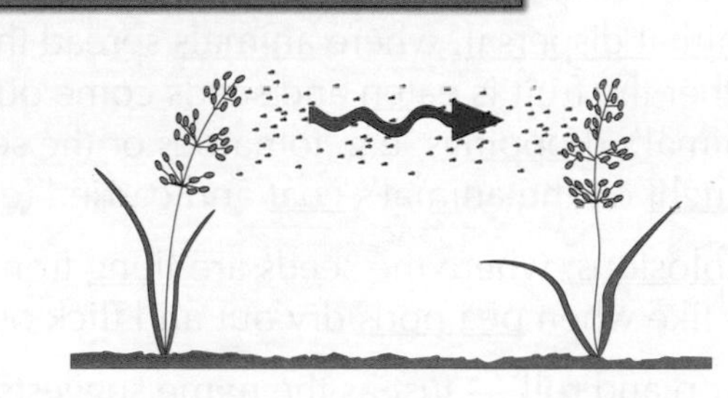

Features of plants that use wind pollination:

1) Usually small dull petals on the flower.
2) No scent or nectaries.
3) Long filaments hang the anthers outside the flower so a lot of pollen is blown away.
4) Stigmas are feathery to catch pollen as it's carried past in the wind.

Fertilisation, Seed Formation and Distribution

Here it is, the follow-up to Plant Reproduction — or what happens after a flower is pollinated.

Fertilisation is the Joining of Sex Cells

1) Pollen is the plant equivalent of human sperm.
2) Pollen grains land on a ripe stigma with help from insects or the wind (see previous page).
3) A pollen tube then grows out of a pollen grain, down through the style to the ovary.
4) The nucleus from a male sex cell moves down the tube to join with a female sex cell inside an ovule. Fertilisation is when the two nuclei join.

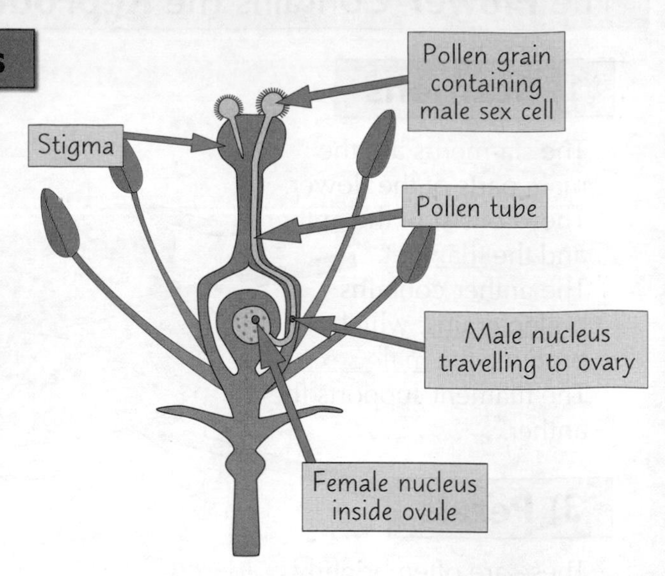

Seeds are Formed from Ovules

1) After fertilisation, the ovule develops into a seed. Each seed contains a dormant (inactive) embryo plant.
2) The embryo plant has a food store which it uses when conditions are right — i.e. when it starts to grow or "germinate".
3) The ovary develops into a fruit around the seed. Fruits can tempt animals to eat them and so scatter the seeds in their faeces ("poo").

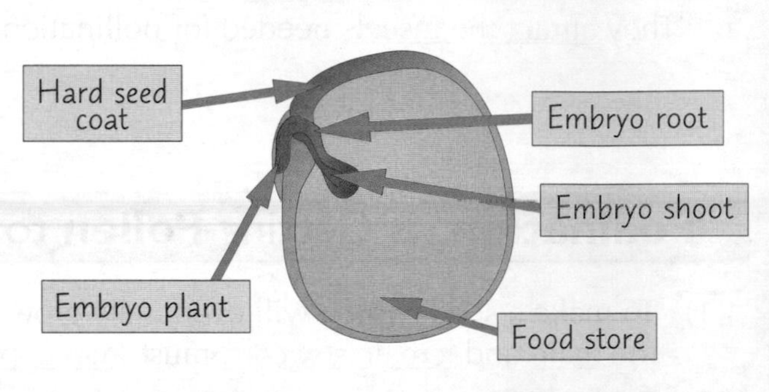

Seed Dispersal is Scattering Seeds

Seeds are dispersed or spread out so that they can grow without too much competition from each other.
Here are some ways in which the seed can be dispersed:

Seeds that have been flung or have rolled away from the parent plant then tend to be further dispersed by animals.

1) Wind dispersal, where the seeds are carried away by the wind, like dandelion or sycamore fruit.
2) Animal dispersal, where animals spread the seeds. Either the fruit is eaten and seeds come out in the animal's droppings, e.g. tomatoes or the seeds get caught on the animal's coat and carried (e.g. burdock fruit).
3) Explosions, where the seeds are flung from the plant — like when pea pods dry out and flick out peas.
4) Drop and roll — just as the name suggests, the fruit falls from the plant and rolls away. An example of this is horse chestnut fruit (conkers).

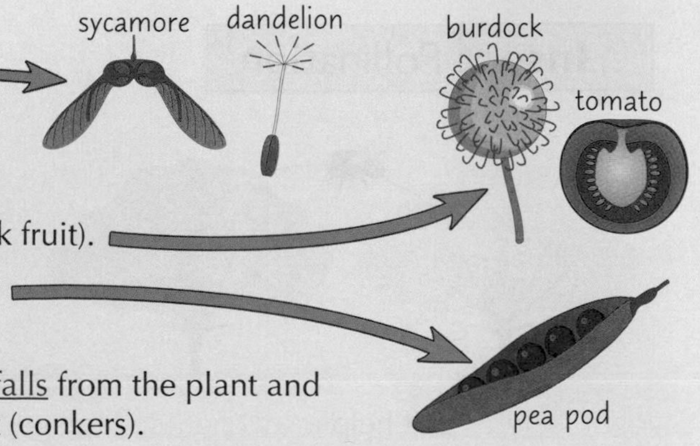

Plants come up with all sorts of ways to disperse their seeds

After pollination and fertilisation, next comes seed development. Then you've got the business of dispersal. Eventually, the seeds will grow into new plants far away from their parents.

Investigating Seed Dispersal Mechanisms

At last, a little bit of science in action. Roll up your sleeves and let's get started.

You Can **Investigate** Seed **Dispersal** by **Dropping Fruit**

You can investigate how well different seeds disperse from the comfort of your own classroom. It's easiest to investigate the wind and drop and roll dispersal mechanisms.

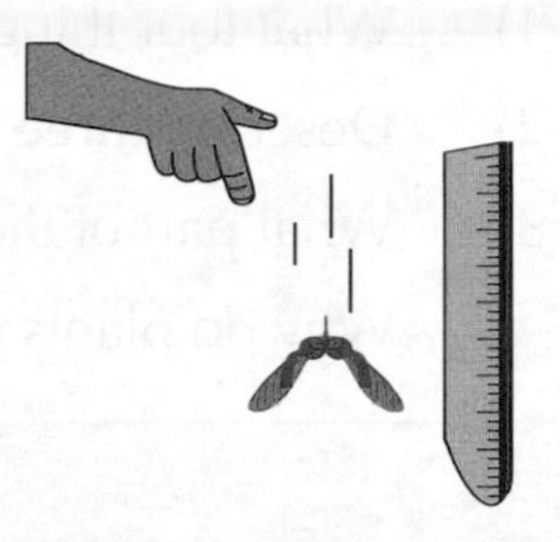

Here's what you have to do.

1) Get yourself some fruit (containing seeds). You could compare ones with different dispersal mechanisms, e.g. sycamore fruit and horse chestnut fruit.
2) Decide on a fixed height to drop the fruit from.
3) Drop the fruit one at a time from this height, directly above a set point on the ground.
4) Using a tape measure, measure and record how far along the ground the seeds have been dispersed.

Do this at least three times for each type of seed and then find the average distance each type travels or 'disperses' when dropped.

Seed Type	Distance Dispersed (cm)		
Sycamore	20	25	
Horse Chestnut			

Make Sure it's a **Fair Test**

So that you can make a fair comparison between the distances travelled by different seed types, you need to keep the following the same each time you do the experiment:

- the person dropping the fruit,
- the height the fruit are dropped from,
- the place you're doing the experiment (stay away from doors and windows that might cause draughts).

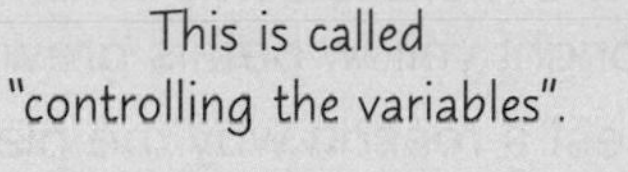

Use a **Fan** to Investigate the "**Wind Factor**"

Many seed dispersal mechanisms are affected by the wind. The special shape of sycamore fruit helps the wind to catch the fruit and carry the seeds far away from the parent sycamore tree.

You can investigate just how much the wind affects seed dispersal by introducing an electric fan into the experiment above.

Here's how:

1) Set up the fan a fixed distance from the person dropping the fruit.
2) Switch the fan on — it needs to be set to the same speed for every fruit you drop. This makes sure the experiment will be a fair test.
3) Drop the fruit as before and measure how far along the ground the seeds travel.

You should find that the sycamore seeds travel much further in windy conditions (i.e. when the fan is switched on). This might not be the case for every seed type though.

WORKING SCIENTIFICALLY

Come on now, fair's fair

Knowing how to control variables to make a test fair is an important part of being a scientist.

Warm-Up and Practice Questions

Photosynthesis and plant reproduction are really important. You're bound to get asked about them at one point or another, so make sure you can answer all these questions without peeking.

Warm-Up Questions

1) What four things are needed for photosynthesis?
2) Describe **three** ways in which leaves are adapted for photosynthesis.
3) What part of the flower develops into the seed after fertilisation?
4) Why do plants disperse their seeds?

Practice Questions

1 Jen found a packet of seeds in her garage. The packet wasn't labelled, so Jen decided to plant the seeds to see what kind of plants grew from them.

(a) Suggest **three** things that will be needed for the seeds to grow into healthy plants.

(3 marks)

(b) After two months some small plants that had flowers with bright yellow petals grew from the seeds. Suggest a reason why the plants had bright yellow petals.

(1 mark)

(c) After the plants had flowered, Jen noticed some seed heads covered in little tiny hooks on the plants. Describe how the hooks would help the plant to disperse its seeds.

(1 mark)

2 The leaves of plants absorb light for photosynthesis.

(a) Write the word equation for photosynthesis using the words below.

oxygen **carbon dioxide** **water** **glucose**

(2 marks)

(b) Rob planted some marigold plants in his garden. He planted some under a tree and some in full sunlight. The plants in full sunlight grew much better than those under the tree. Suggest why the plants grew better in full sunlight.

(2 marks)

(c) Rob also planted some marigold plants in his greenhouse. Half the marigolds were planted in mineral-rich compost bought in a shop. The other half were planted in ordinary soil from the garden. Suggest which group of marigolds were healthier. Explain your answer.

(2 marks)

Practice Questions

3 Elspeth is investigating how well different seeds are dispersed by wind. She sets up a fan and drops a sycamore fruit and a horse chestnut fruit in front of it. She then measures how far along the ground each of them travels. She does this three times for each fruit.

(a) Give **two** things Elspeth needs to keep the same each time she repeats the experiment to make sure it is a fair test.

(2 marks)

(b) Which fruit would you expect to disperse its seed(s) further? Explain your answer.

(2 marks)

4 (a) Chris and Jim were talking about fertilisation in plants.
Chris said that "fertilisation in plants happens when the pollen grain lands on the stigma".
Jim said that "fertilisation is when the nuclei from the ovule (egg) and male sex cell actually join together".
Who is correct, Jim or Chris?

(1 mark)

(b) The sentences below can be rearranged to describe the stages that must occur if a plant is to reproduce successfully. Copy and complete the table, numbering the steps 1 to 5 to show the correct order of these events.
The first one has been done for you.

stage	order
The nucleus of the male sex cell joins with the nucleus of the egg cell (ovule).	4
Pollen grain lands on the stigma.	1
The ovary develops into a fruit with the seeds inside.	5
The nucleus from a male sex cell moves down through the tube.	3
A pollen tube grows down through the style to the ovary.	2

(4 marks)

Dependence on Other Organisms

Organisms in an **Ecosystem** are **Interdependent**

1) An ecosystem is all the living organisms in one area, plus their environment.
2) The organisms in an ecosystem are interdependent — they need each other to survive.

Almost **All Living Things Depend** on **Plants**

Almost all life on Earth depends on plants. Without them, we just wouldn't be here. Here's why...

Plants **Capture** the **Sun's Energy**

1) Almost all energy on Earth comes from the Sun.
2) Plants use some of the Sun's energy to make food during photosynthesis (see page 30). They then use this food to build "organic molecules" (things like carbohydrates and proteins), which become part of the plants' cells.
3) These organic molecules store the Sun's energy. The energy gets passed on from plants to animals when animals eat the plants. It gets passed on again when these animals are eaten by other animals.
4) Only plants, algae (seaweeds) and some bacteria are able to carry out photosynthesis. So nearly all living things rely on plants to capture and store the Sun's energy.

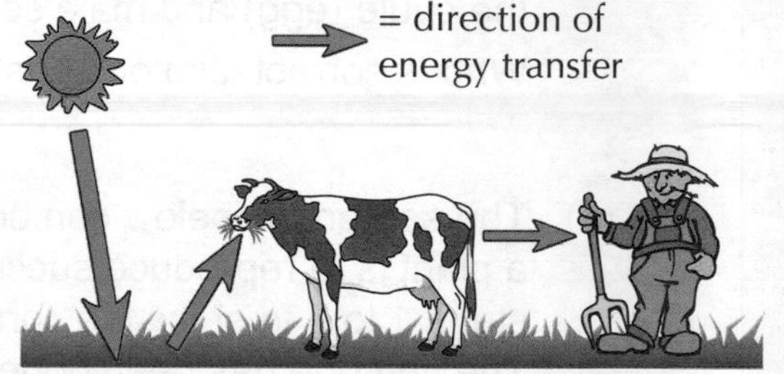

Plants **Release Oxygen** and **Take in Carbon Dioxide**

1) All living things respire (see page 4).
2) When plants and animals respire, they take in oxygen (O_2) from the atmosphere and release carbon dioxide (CO_2).
3) When plants photosynthesise, they do the opposite — they release oxygen and take in carbon dioxide.
4) So photosynthesis helps make sure there's always plenty of oxygen around for respiration. It also helps to stop the carbon dioxide level in the atmosphere from getting too high. This is an example of organisms affecting their environment.

Many **Plants** Depend on **Insects** in Order to **Reproduce**

1) Many plants depend on insects to pollinate them (see page 31).
2) Without insects like bees, moths and butterflies, these plants would struggle to reproduce.
3) This would obviously be bad for the plants, but it would be bad for humans too. Many of our crop plants need to be pollinated by insects in order to produce the fruit, nuts and seeds that we eat.
4) So we depend on insects to pollinate our crops and ensure our food supply.

Food Chains

Organisms depend on each other to survive. Mainly this means that they depend on each other for food.

Food Chains Show What is Eaten by What

1) The organisms in a food chain are usually in the same ecosystem.

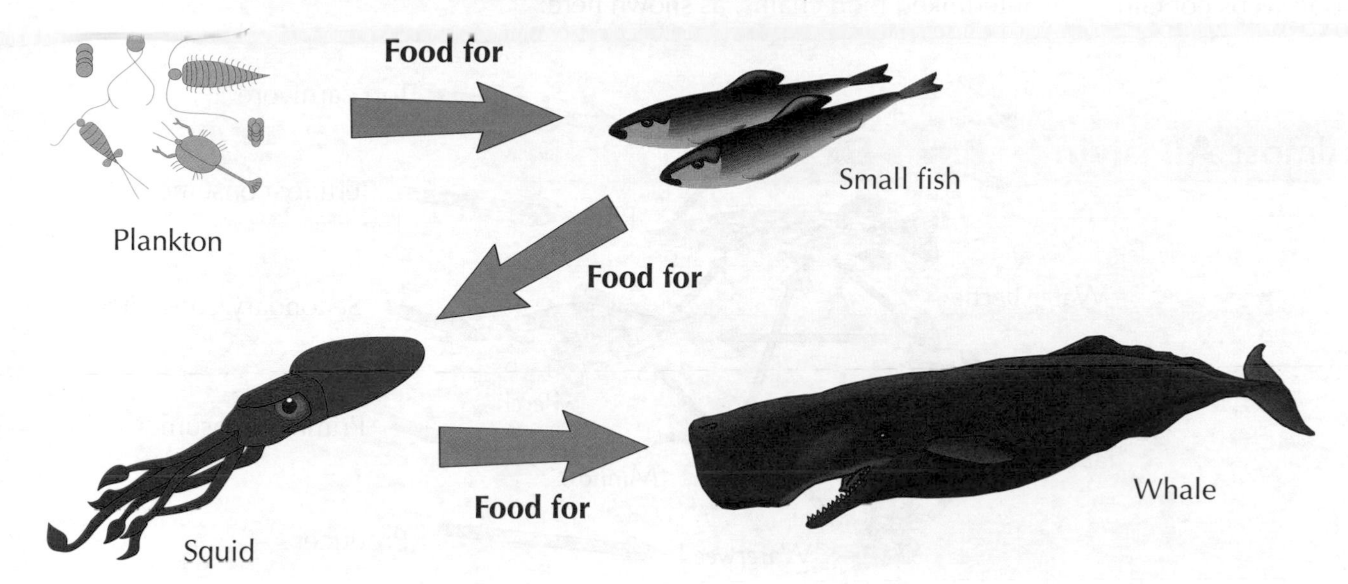

2) The arrows show what is eaten by what — i.e. "food for". (Plankton is food for small fish, etc.)
3) The arrows also show the direction of energy flow.

Poisons Build Up as They are Passed Along a Food Chain

Toxic materials (poisons) can sometimes get into food chains and harm the organisms involved. Organisms higher up the food chain (usually the top carnivores) are likely to be the worst affected as the toxins accumulate (build up) as they are passed along.

A top carnivore is an organism that isn't eaten by anything else.

Crops

Small birds

Bird of Prey

■ = level of poison

Food chains show what's highest in the pecking order

Food chains are simple, so you've no excuse not to learn them. They show you what eats what, right up to the top carnivore. In the exam, if you're asked to draw a food chain, make sure you have the arrows pointing the correct way — you don't want to say a leaf eats a snail.

Food Webs

You saw simple food chains on the last page — now it's time to delve into the more complicated food webs. In a food web, lots of the animals and plants are linked together in multiple ways.

Food Webs and Their Tremendous Terminology

Food webs contain many interlinked food chains, as shown here:

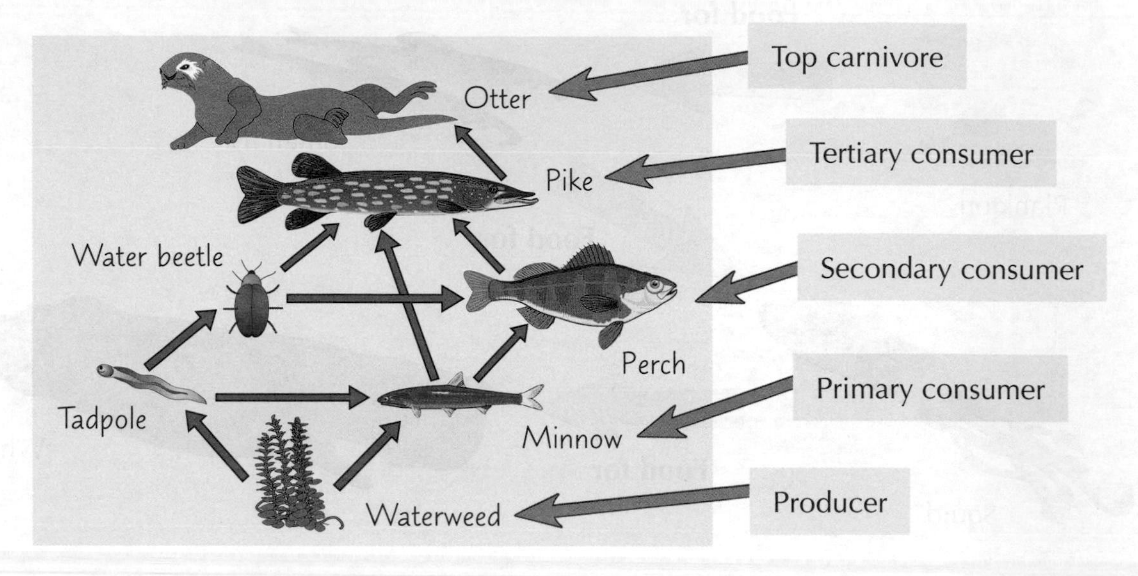

Learn these nine bits of terminology:

1) PRODUCER — all plants are producers. They use the Sun's energy to 'produce' food energy.
2) HERBIVORE — an animal that only eats plants, e.g. tadpoles, rabbits, caterpillars, aphids.
3) CONSUMER — all animals are consumers. (All plants are not, because they're producers.)
4) PRIMARY CONSUMER — an animal that eats producers (plants).
5) SECONDARY CONSUMER — an animal that eats primary consumers.
6) TERTIARY CONSUMER — an animal that eats secondary consumers.
7) CARNIVORE — eats only animals, never plants.
8) TOP CARNIVORE — is not eaten by anything else.
9) OMNIVORE — eats both plants and animals.

The organisms in a food web are all interdependent — so a change in one organism can easily affect others.

Example — What happens if the minnows are removed?
1) Who will get eaten LESS? The tadpoles, as there are no minnows there to eat them.
2) Who will get eaten MORE? a) Water beetles (by perch who'll get hungry without minnows).
b) Waterweeds (since the numbers of tadpoles will increase).

Learn about food webs — but don't get tangled up

Once you've got this page learnt, you can practise this typical food web question: "If the number of otters decreased, give one reason why the number of water beetles might a) decrease b) increase". You can find the answer to this question on page 190.

Warm-Up and Practice Questions

It's easy to think you've learnt everything in the section until you try to answer the Warm-Up Questions. If that happens to you, don't worry, just go back over the pages and write out the bits you got wrong until you can answer them standing on your head.
Then stand on your head and try to answer the Practice Question.

Warm-Up Questions

1) All the organisms in an ecosystem are interdependent. What does this mean?
2) Explain why animals rely on plants for energy.
3) What do the arrows in a food chain or web represent?
4) Draw the food chain for plankton, squid, small fish and whales.
5) Give a definition for each of the following food web terminologies:
   a) producer   b) carnivore   c) omnivore

Practice Question

1 Below is part of the food web of plants and animals in the Arctic.

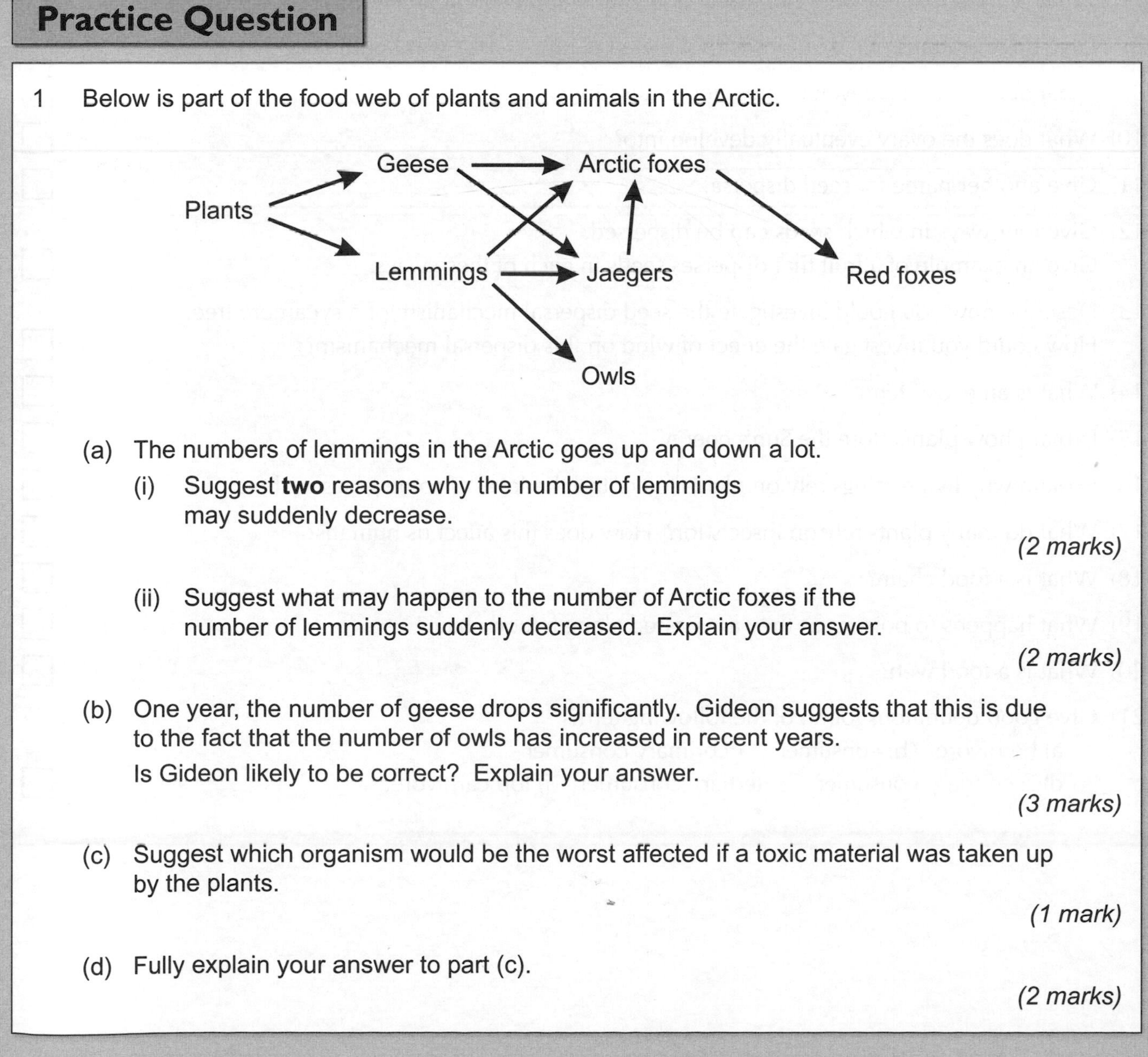

(a) The numbers of lemmings in the Arctic goes up and down a lot.

(i) Suggest **two** reasons why the number of lemmings may suddenly decrease.

(2 marks)

(ii) Suggest what may happen to the number of Arctic foxes if the number of lemmings suddenly decreased. Explain your answer.

(2 marks)

(b) One year, the number of geese drops significantly. Gideon suggests that this is due to the fact that the number of owls has increased in recent years.
Is Gideon likely to be correct? Explain your answer.

(3 marks)

(c) Suggest which organism would be the worst affected if a toxic material was taken up by the plants.

(1 mark)

(d) Fully explain your answer to part (c).

(2 marks)

Revision Summary for Section Three

Green plants are ace aren't they? What I really like about them is that they're all so clean and fresh — human and animal biology always seems to end up so gory with all sorts of gruesome diagrams and horrid diseases. But plants have such simple lives — they just seem to "go with the flow", with no apparent discomfort and no worries — and let's face it, it's a nice trick if you can do it.

Alas, nature conspired to give humans an altogether more "challenging" experience on this little blue-green planet of ours — and somehow that's ended up with you needing to know the answers to all these questions. Hmmm, it's a funny old world isn't it — when you think about it from that angle... Anyway, here they are. Off you go then...

1) What is made during photosynthesis?
2) What do plants do with glucose?
3) What is the by-product made in photosynthesis, which is needed by animals?
4) Apart from water, what do plants need from the soil?
5) What are the four main parts of a flower? Say what each part actually does.
6) What is pollination? What are the two types of pollination?
7) What is the difference between insect pollination and wind pollination?
8) Give three features of: a) an insect pollinated plant, b) a wind pollinated plant.
9) What does an ovule develop into after fertilisation?
10) What does the ovary eventually develop into?
11) Give another name for seed dispersal.
12) Give four ways in which seeds can be dispersed.
Give an example of a fruit that disperses seeds in each of these ways.
13) Describe how you could investigate the seed dispersal mechanism of a sycamore tree.
How could you investigate the effect of wind on this dispersal mechanism?
14) What is an ecosystem?
15) Explain how plants store the Sun's energy.
16) Explain why living things rely on plants to control the level of some gases in the air.
17) What do many plants rely on insects for? How does this affect us humans?
18) What is a food chain?
19) What happens to poisons as they are passed along a food chain?
20) What is a food web?
21) Give good definitions for all of the following terms:
a) herbivore b) consumer c) primary consumer
d) secondary consumer e) tertiary consumer f) top carnivore.

DNA and Inheritance

DNA is a bit like your body's own instruction manual. When you're being made, you get bits of DNA from your mum and bits from your dad — this is how you inherit characteristics.

Chromosomes, DNA and Genes

1) Most cells in your body have a nucleus. The nucleus contains chromosomes.
2) Chromosomes are long, coiled up lengths of a molecule called DNA.
3) DNA is a long list of chemical instructions on how to build an organism.
4) A gene is a short section of a chromosome (and so a short section of DNA).
5) Genes control many of our characteristics, e.g. hair colour, eye colour, hairiness, etc. Different genes control different characteristics.
6) Genes work in pairs — one will usually be dominant over the other.

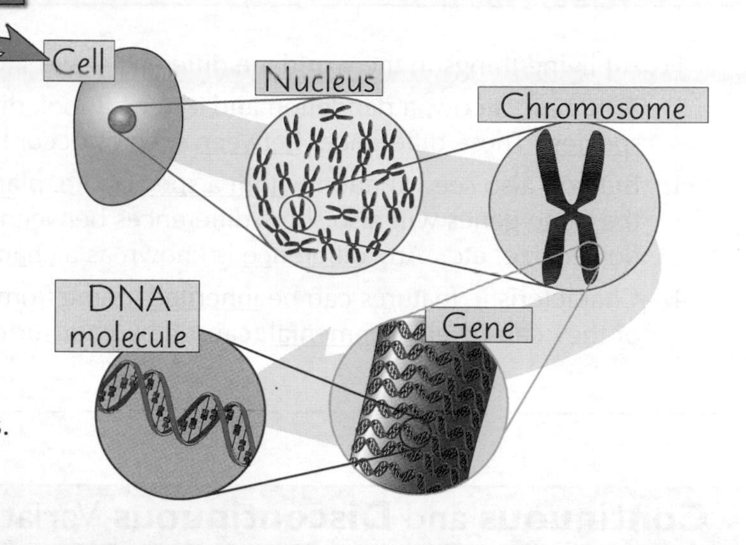

We Inherit Characteristics from Our Parents

1) Human body cells have 46 chromosomes (23 pairs).
2) Sperm and egg cells carry only 23 chromosomes.

3) During reproduction, when an egg is fertilised, the nucleus of the egg fuses with the nucleus of the sperm.
4) This means that the fertilised egg contains 23 matched pairs of chromosomes. It has one copy of each gene from the mother and one from the father.
5) Since genes control characteristics, the fertilised egg develops into an embryo with a mixture of the parents' characteristics. This is how you 'inherit' your parents' characteristics.
6) The process by which genes are passed down from parents to their offspring is called heredity.

A characteristic passed on in this way is called a 'hereditary' characteristic.

The Structure of DNA Was Only Worked Out Recently

1) Scientists struggled for decades to work out the structure of DNA.
2) Crick and Watson were the first scientists to build a model of DNA — they did it in 1953.
3) They used data from other scientists, Wilkins and Franklin, to help them understand the structure of the molecule. This included X-ray data showing that DNA is a double helix — a spiral made of two chains wound together.
4) By putting all the information together, Crick and Watson were able to build a model showing what DNA looks like.

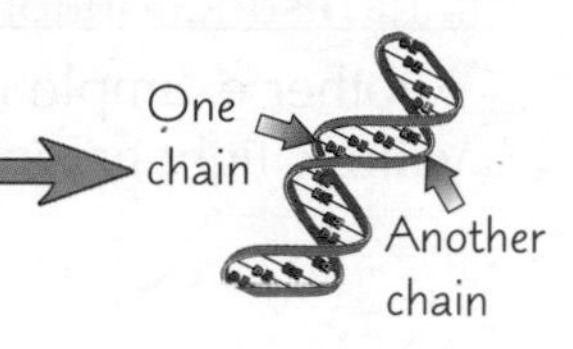

Some important details to learn on this page

There are three main headings, sixteen numbered points and two important diagrams — and they all need learning. Sit down and challenge yourself to repeat the main details. If you struggle with any bits, reread the page, then cover it back up and try again.

Variation

This page is all about differences between organisms — both big, obvious differences, like those between a tree and a cow, and less obvious differences, like people having different blood groups.

Different Species Have Different Genes

1) All living things in the world are different — we say that they show VARIATION.
2) A human, a cow, a dandelion and a tree all look different because they're different species. These differences between species occur because their genes are very different.
3) But you also see variation within a species, i.e. plants or animals that have basically the same genes will also show differences between them, e.g. skin colour, height, flower size, etc. Any difference is known as a characteristic feature.
4) Characteristic features can be inherited (come from your parents via genes) or they can be environmental (caused by your surroundings).

Continuous and Discontinuous Variation

Variation between individuals within a species can either be classed as continuous or discontinuous.

Continuous Variation — the feature can vary over a range of values

1) Examples of this are things like height, weight, skin colour, intelligence, leaf area, etc. where the feature can have any value at all — within a certain range.
2) If you did a survey of kids' heights you could plot the results on a chart like the one to the right (the heights would be collected into groups to give the bars).
3) The smooth distribution curve drawn on afterwards (the red line) shows much better the continuous way that values for height actually vary.

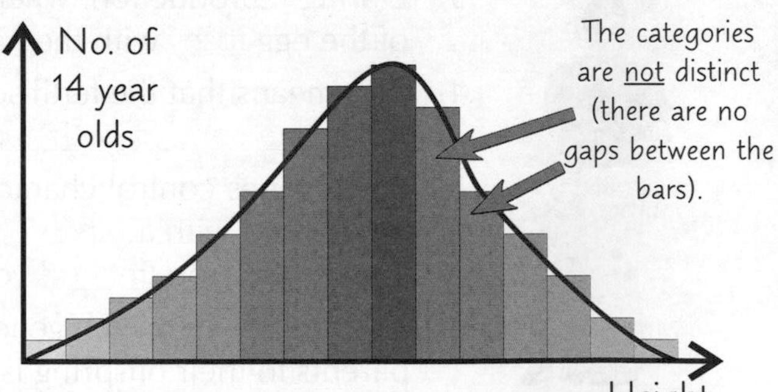

Discontinuous Variation — the feature can only take certain values

1) An example of this is a person's blood group, where there are just four distinct options, NOT a whole continuous range.
2) Another example is the colour of a courgette. A courgette is either yellow, light green or dark green — there's no range of values.

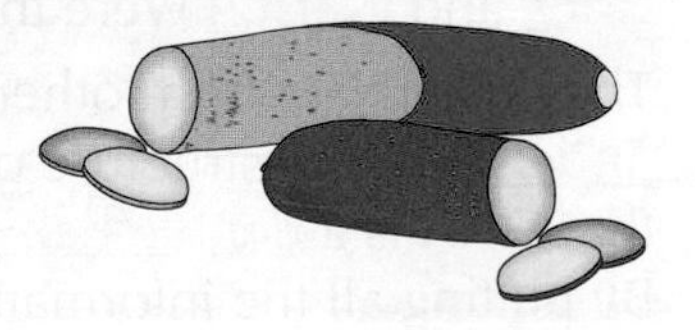

You need to be able to explain variation in terms of genes

Don't let the fancy word "variation" put you off. It's really not as complicated as it sounds. It just means "differences" (between any living things). You can have variation (differences) between different species, and you can also have variation (differences) within one species.

Natural Selection and Survival

In nature, being different can be really important. Having a different characteristic to other organisms can determine whether an organism (and its future generations) is likely to survive in the long run or not.

Variation Leads to Natural Selection

1) Organisms show variation because of differences in their genes (see previous page).
2) Organisms also have to compete for the resources they need in order to survive and reproduce, e.g. food, water and shelter. They have to compete with other members of their own species, as well as organisms from other species.

FOR EXAMPLE...

...this red squirrel...

...has to compete with other red squirrels (members of its own species)...

...as well as grey squirrels (a different species), in order to get food.

3) Organisms with characteristics that make them better at competing are more likely to survive and reproduce. This means they're more likely to pass on the genes for their useful characteristics to the next generation.
4) Organisms that are less successful competitors are usually the first to die — possibly before they've had a chance to reproduce. This means their genes and less useful characteristics won't be passed on to any offspring.
5) So, over time, the gene for a useful characteristic will become more common.
6) This process in which a characteristic gradually becomes more (or less) common in a population is known as natural selection.

A population is all the organisms of one species that live in the same ecosystem.

Giraffes Have Long Necks Due to Natural Selection

Are you sitting comfortably...

Once upon a time there was a group of animals munching leaves from a tree. Unfortunately the population was high and food was running short. Soon all the leaves on the lower parts of the trees were gone and the animals started to get hungry — some even died. Except, that is, for a couple of animals which happened to have slightly longer necks than normal. This meant that they could compete better for food — they could reach just that bit higher, to the juicy and yummy leaves higher up the trees.

They survived that year, unlike a lot of animals, and had lots of babies.

The babies also had longer necks, and could eventually reach up the tree for the juicy yummy leaves.

It soon got to a situation where most of the animals in the population had long necks...

It's all about competition and being the best

Only those who are born with features that make them great at competing in the world they live in are likely to survive and produce offspring — the sick and the inept all die off very quickly.

Extinction and Preserving Species

Organisms that can't compete don't survive for long. If they suddenly become less competitive due to changes in the environment, they could die out in a certain area — or even become extinct.

Many Species Are at Risk of Becoming **Extinct**

1) Many organisms survive because they are well-adapted for competing in their environment.
2) But if the environment changes in some way, some organisms may struggle to compete successfully for the resources they need to survive and reproduce.
3) If this happens to a whole species, then that species is at risk of becoming extinct. Extinct means that there are none of them left at all (like the woolly mammoth).
4) Species at risk of becoming extinct are called endangered species.

Humans Can **Suffer** When Species Become **Extinct**

There are probably loads of species we don't know about, e.g. in unexplored rainforests and deep in the ocean.

1) Humans rely on plants and animals for food.
2) We also use them to make clothing, medicines, fuel, etc.
3) We need to protect the organisms we already use in this way. We also need to make sure organisms we haven't discovered yet don't become extinct before we find them — or we might miss out on new sources of useful products.
4) Ecosystems are complex. If one species becomes extinct, this can have a knock-on effect for other organisms — including us.
5) That's why it's important for us to maintain the planet's biodiversity — the variety of species that live on Earth.

Gene Banks May Help to **Prevent Extinction**

1) A gene bank is basically a store of the genes of different species.
2) This means that if a species becomes endangered or even extinct, it may be possible to create new members of that species. So gene banks could be a way of maintaining biodiversity in the future.
3) Genes are stored differently for plants and animals. For example:

Animals

Sperm and egg cells (which contain genes) may be frozen and stored.
Scientists could then use these cells to create new animal embryos in the future.

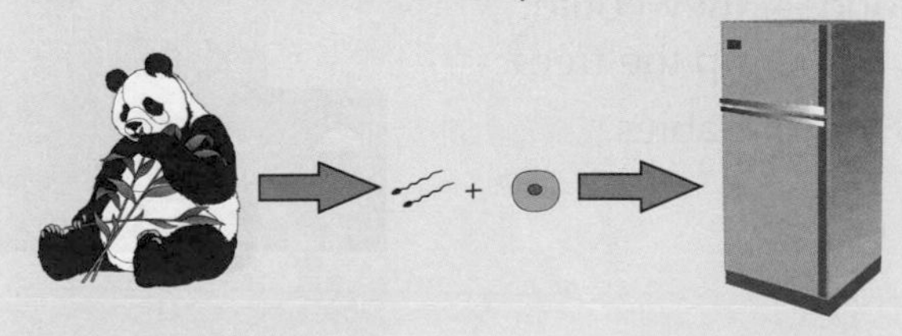

Plants

Seeds (which contain genes) can be collected from plants and stored in seed banks. If the plants become extinct in the wild, new plants can be grown from the seeds kept in storage.

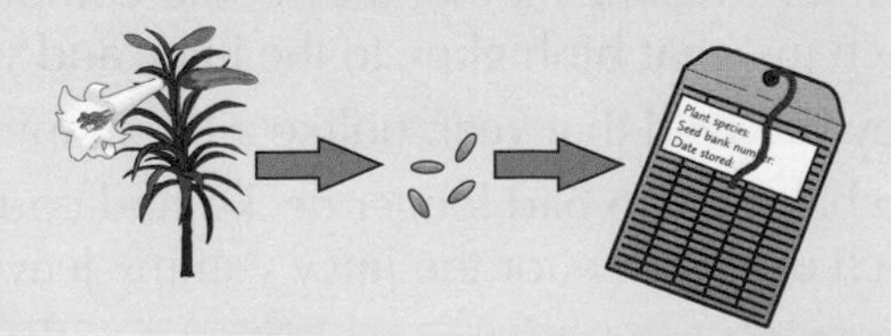

Gene banks aren't the only way to maintain biodiversity. It's much better to try to stop species becoming extinct in the first place, e.g. by preventing the destruction of habitats (the areas where organisms live).

It's not just animals that suffer when they go extinct

Underline key words in exam questions to make sure your answer covers exactly what the question asks — it's no good telling them all about gene banks if you won't get marks for it.

Warm-Up and Practice Questions

If you don't take time to warm up you're risking some serious brain-strain. So take a look at these quick questions and get your mind nice and supple on some of the basic facts.
Then launch yourself slowly into the exam questions and enjoy.

Warm-Up Questions

1) What are chromosomes?
2) What did Watson and Crick build in 1953?
3) Can you get variation within a species?
4) What two things can cause characteristic features?
5) What are the two different classes of variation? How are they different?
6) How can scientists help to maintain biodiversity?
7) What's a habitat? Why is preserving habitats important to keeping species alive?

Practice Questions

1 (a) Copy and complete the labels in the diagram below

(i) DNA....... (ii) Gene...... (iii) Chromosome

(3 marks)

(b) The double helix model describing the structure of DNA was first developed in 1953.

(i) What is DNA?

(1 mark)

(ii) What is a double helix?

(1 mark)

(c) A gene is a short section of DNA. Genes are found on chromosomes.

(i) Give three examples of characteristics controlled by genes.

(3 marks)

(ii) How many matched pairs of chromosomes does a fertilised human egg contain?

(1 mark)

(iii) What is the name of the process that describes how genes are passed on from parents to their offspring?

(1 mark)

Practice Questions

2 Kate wants to get a pet rabbit. She looks at several rabbits in the pet shop and notices that some have long, straight ears and some have large, floppy ears even though they all belong to the same species.

(a) (i) Explain why the rabbits can have different types of ear even though they belong to the same species.

(1 mark)

(ii) Is this variation in ear type continuous or discontinuous? Explain your answer.

(2 marks)

(b) The picture below shows a typical rabbit. Suggest why rabbits have evolved to have big ears.

(3 marks)

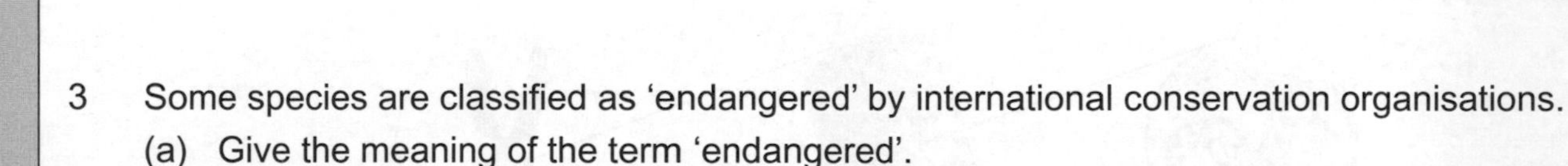

3 Some species are classified as 'endangered' by international conservation organisations.

(a) Give the meaning of the term 'endangered'.

(1 mark)

(b) Explain how a change in the environment could lead to a species becoming endangered.

(3 marks)

(c) Explain why it is important to preserve endangered plant and animal species.

(3 marks)

(d) One way to maintain diversity is to store the genes of endangered species in gene banks. Explain how genes can be stored in and retrieved from gene banks for:

(i) endangered plants

(2 marks)

(ii) endangered animals

(2 marks)

Revision Summary for Section Four

Section Four is fairly basic stuff really, but there are one or two fancy words which might cause you quite a bit of grief until you've made the effort to learn exactly what they mean: "DNA" is just a list of instructions for how any living creature is put together; "variation" just means "differences", etc., etc.

These questions aren't the easiest you could find, but they test exactly what you know and find out exactly what you don't. You need to be able to answer them all, because all they do is test the basic facts. You must practise these questions over and over again until you can just sail through them.

1) Where do you find chromosomes?
2) What are chromosomes made of?
3) What is a gene? What do genes control?
4) How many chromosomes do humans have in each body cell?
5) How many chromosomes are there in human sperm cells? How about in human egg cells?
6) What happens at fertilisation?
7) What does heredity mean?
8) Name the two scientists who first built a model of DNA. Name the other two scientists whose data helped them.
9) Describe the structure of a DNA molecule.
10) What does variation mean?
11) Why do different species look different?
12) What is a characteristic feature of an organism?
13) What is continuous variation? Give three examples.
14) What is discontinuous variation? Give two examples.
15) Give one way in which a graph showing continuous variation would differ from a graph showing discontinuous variation.
16) Why is it important that organisms are good at competing for the things they need?
17) Why are genes for useful characteristics likely to become more common in a population over time? What is this process called?
18) How did giraffes end up with very long necks?
19) Why could it be bad news for an organism if its environment changes?
20) What does extinct mean?
21) What does endangered mean?
22) What is biodiversity? Why is it important for us to maintain the planet's biodiversity?
23) What is a gene bank? What are they used for?
24) What part of a plant may be stored in a gene bank? What about an animal?

Solids, Liquids and Gases

The first page in this section is all about states of matter and there are only three you need to know.

The Three States of Matter — Solid, Liquid and Gas

1) Materials come in three different forms — solids, liquids and gases.
2) These are called the Three States of Matter.
3) All materials are made up of tiny particles.
4) Which state you get (solid, liquid or gas) depends on how strongly the particles stick together. How well they stick together depends on three things:
   a) the material b) the temperature c) the pressure.

Solids, Liquids and Gases Have Different Properties

1) We can recognise solids, liquids and gases by their different properties.
2) A property of a substance is just a way of saying how it behaves.

Property	Solids	Liquids	Gases
Volume This is how much space something takes up.	Solids have a definite volume	Liquids have a definite volume	Gases have no definite volume — they always fill the container they're in
Shape	Solids have a definite shape	Liquids match the shape of the container	Gases become the same shape as the container
Density This is how heavy something is for its size.	Solids usually have a high density (heavy for their size)	Liquids usually have medium density	Gases have a very low density
Compressibility This is how much you can squash something.	Solids are not easily squashed	Liquids are not easily squashed	Gases are easily squashed
Ease of Flow	Solids don't flow	Liquids flow easily	Gases flow easily

Particle Theory

Particle theory — sounds pretty fancy. But actually it's pretty straightforward.

It's all about the **Energy** and **Arrangement** of **Particles**

1) The particles in a substance stay the same whether it's a solid, a liquid or a gas.
2) What changes is the arrangement of the particles and their energy.

Solids — Particles are Held **Very Tightly Together**

1) The particles in a solid have the least energy.
2) There are strong forces of attraction between particles.
3) The particles are held closely in fixed positions in a very regular arrangement. But they do vibrate to and fro.
4) The particles don't move from their positions, so all solids keep a definite shape and volume, and can't flow like liquids.
5) Solids can't easily be compressed because the particles are already packed very closely together.
6) Solids are usually dense, as there are lots of particles in a small volume.

Liquids — Particles are **Close Together** But They Can **Move**

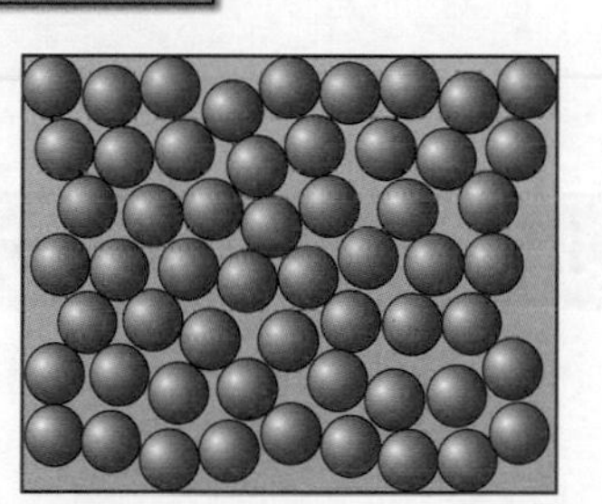

1) The particles in a liquid have more energy.
2) There are some forces of attraction between the particles.
3) The particles are close, but free to move past each other — and they do stick together. The particles are constantly moving in all directions.
4) Liquids don't keep a definite shape and can form puddles. They flow and fill the bottom of a container. But they do keep the same volume.
5) Liquids won't compress easily because the particles are packed closely together.
6) Liquids are quite dense, as there are quite a lot of particles in a small volume.

Gases — Particles are **Far Apart** and **Whizz About a Lot**

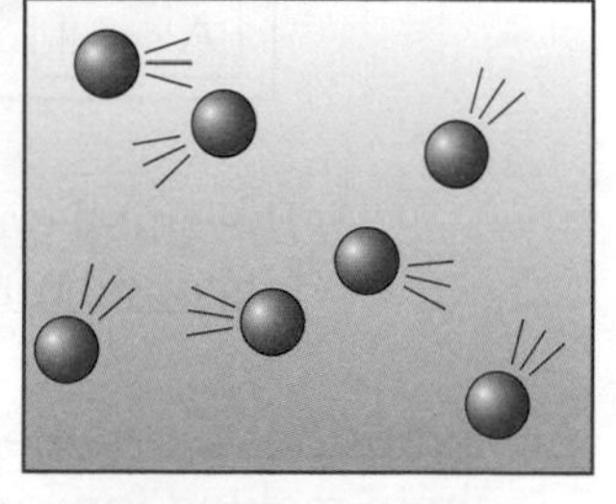

1) The particles in a gas have the most energy.
2) There are very weak forces of attraction between the particles.
3) The particles are far apart and free to move quickly in all directions.
4) The particles move fast, and so collide with each other and the container.
5) Gases don't keep a definite shape or volume and will always expand to fill any container. Gases can be compressed easily because there's a lot of free space between the particles.
6) Gases all have very low densities, because there are not many particles in a large volume.

The particles in gases are far apart and have lots of energy

I think it's pretty clever the way you can explain all the differences between solids, liquids and gases with just a page full of blue snooker balls. Anyway, that's the easy bit. The not-so-easy bit is making sure you've learnt it all. Keep at it and you'll get to grips with what the particles are up to in no time.

More Particle Theory

Gas Pressure is Due to Particles Hitting a Surface

Increasing the Temperature Increases Pressure

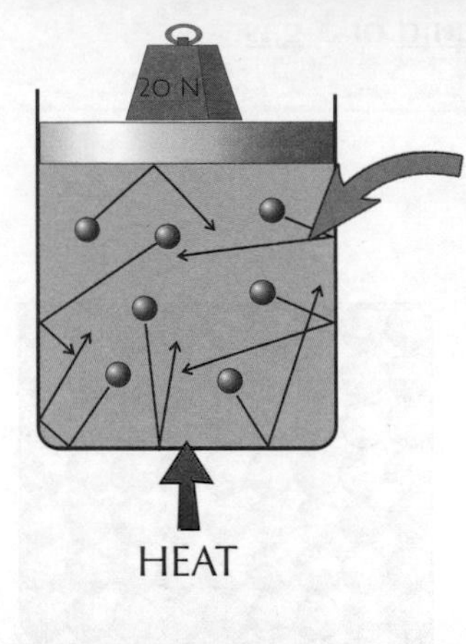

1) When you increase the temperature, it makes the particles move faster.
2) This has two effects:

   a) They hit the walls harder.
   b) They hit more often.

Increasing the temperature will only increase the pressure if the volume stays the same (and vice versa).

3) Both these things increase the pressure.

Reducing the Volume Increases Pressure

1) If you reduce the volume it makes the pressure increase.
2) This is because when the particles are squashed up into a smaller space they'll hit the walls more often.

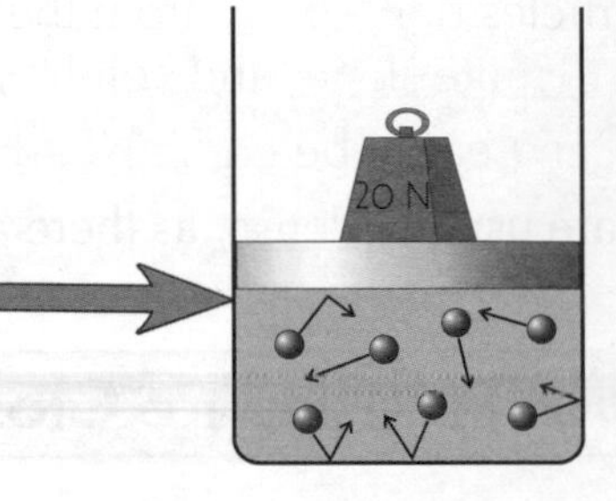

Diffusion is Just Particles Spreading Out

1) Particles "want" to spread out — this is called diffusion. An example is when a smell spreads slowly through a room.

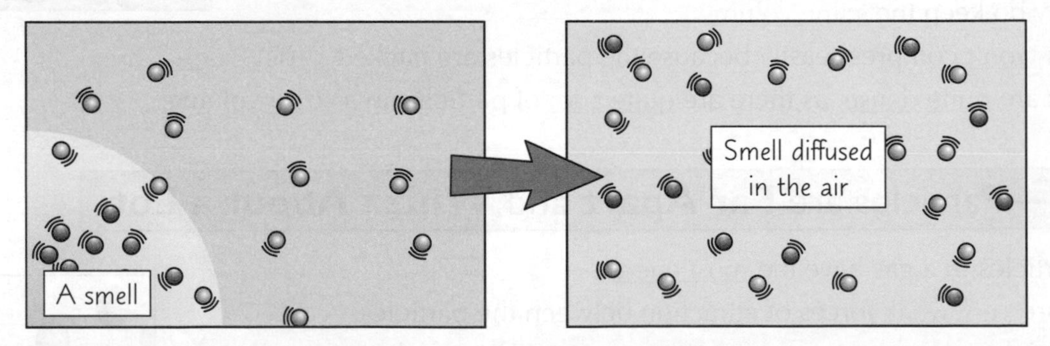

The smell particles move from an area of high concentration (i.e. where there are lots of them) to an area of low concentration (where there's only a few of them).

2) Diffusion is slow because the smell particles keep bumping into air particles, which stops them making forward progress and often sends them off in a completely different direction.

Think about gases squashed in an aerosol can

Aerosols hold gases under pressure, and when you spray an aerosol, you get to smell diffusion in action. Marvellous. Now cover the page and see how much you can write down.

Physical Changes

Physical changes don't change the particles — just their arrangement or their energy.

Physical Changes can be **Changes of State**

— i.e. changing from one state of matter to another.

3) At a certain temperature, the particles have enough energy to break free from their positions. This is called melting and the solid turns into a liquid.

4) When a liquid is heated, again the particles get even more energy.

5) This energy makes the particles move faster which weakens the forces holding the liquid together.

2) This makes the particles move more which weakens the forces that hold the solid together. This makes the solid expand.

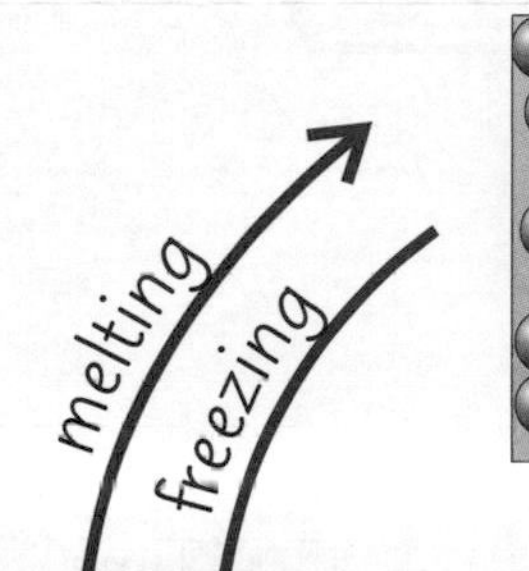

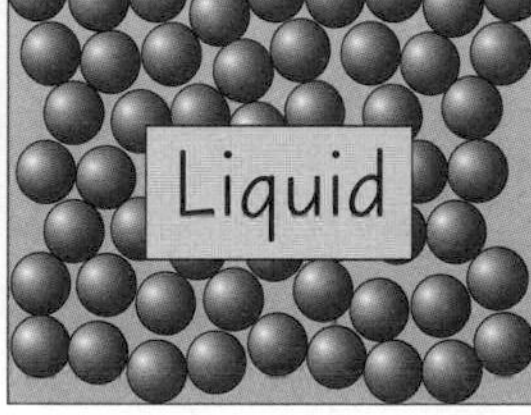

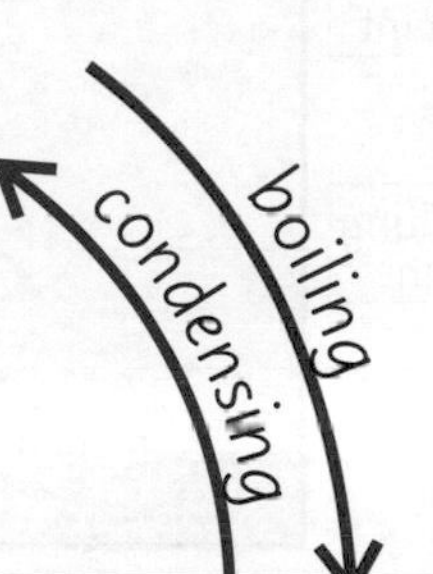

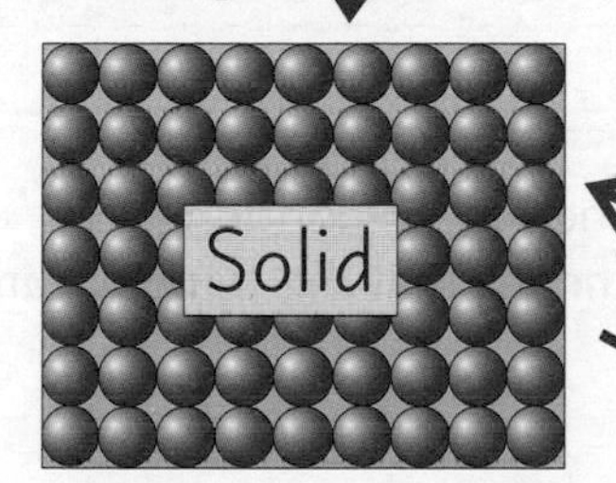

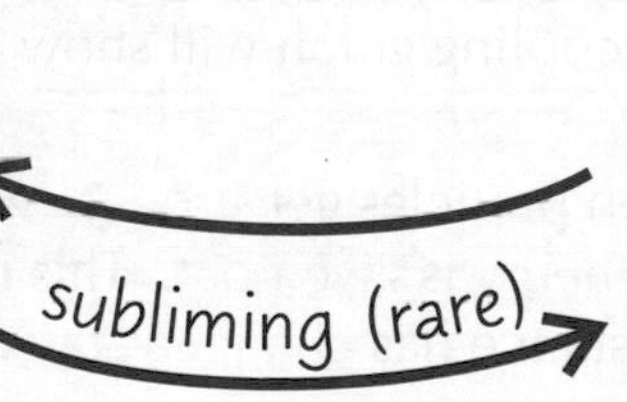

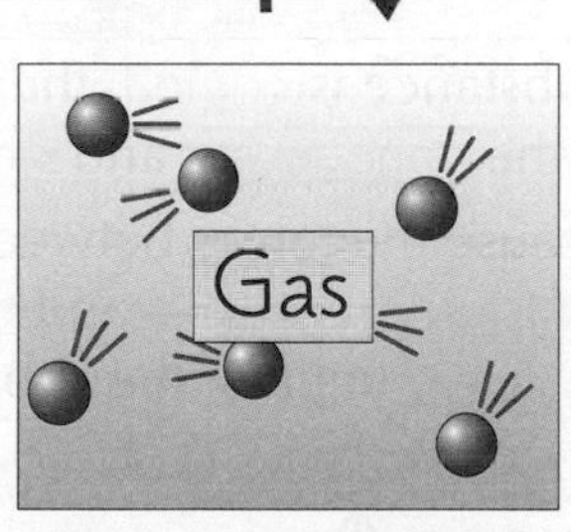

1) When a solid is heated, its particles gain more energy.

6) At a certain temperature, the particles have enough energy to break free of the forces. This is called boiling and the liquid turns into a gas.

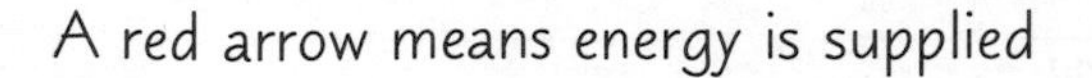

A red arrow means energy is supplied

A blue arrow means energy is given out

A change of state doesn't involve a change in mass, only a change in energy.

From the sublime to the ridiculous

So matter can move from solid to liquid to gas and back again. Learn this, and learn what happens when it changes state. Write it all down bit by bit. Start with the diagram, then add the five labels — then try to learn all the details that go with each one. You'll know it all in no time — you'll see.

Heating and Cooling Curves

When a substance changes state, its temperature stops increasing or decreasing for a while.

Heating and Cooling Curves have Flat Bits

Heating and cooling curves show the energy changes that happen when a substance changes state.

1) When a substance is melting or boiling, all the energy supplied from heating is used to weaken the forces between particles rather than raising the temperature — hence the flat bits on the heating graph.

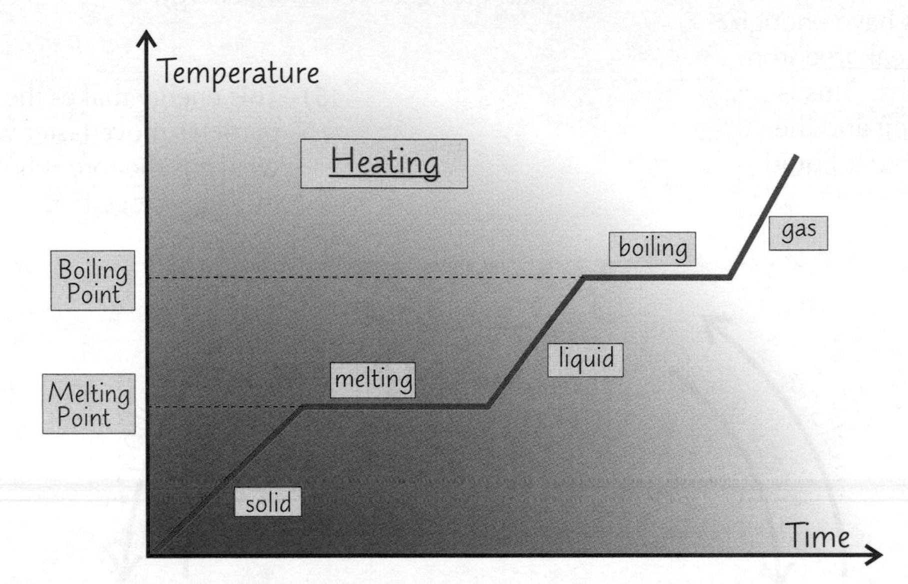

2) When a substance is cooled, the cooling graph will show flat bits at the condensing and freezing points.
3) This is because the forces between particles get stronger when a gas condenses or when a liquid freezes — and energy is given out. This means that the temperature doesn't go down until all the substance has changed state.

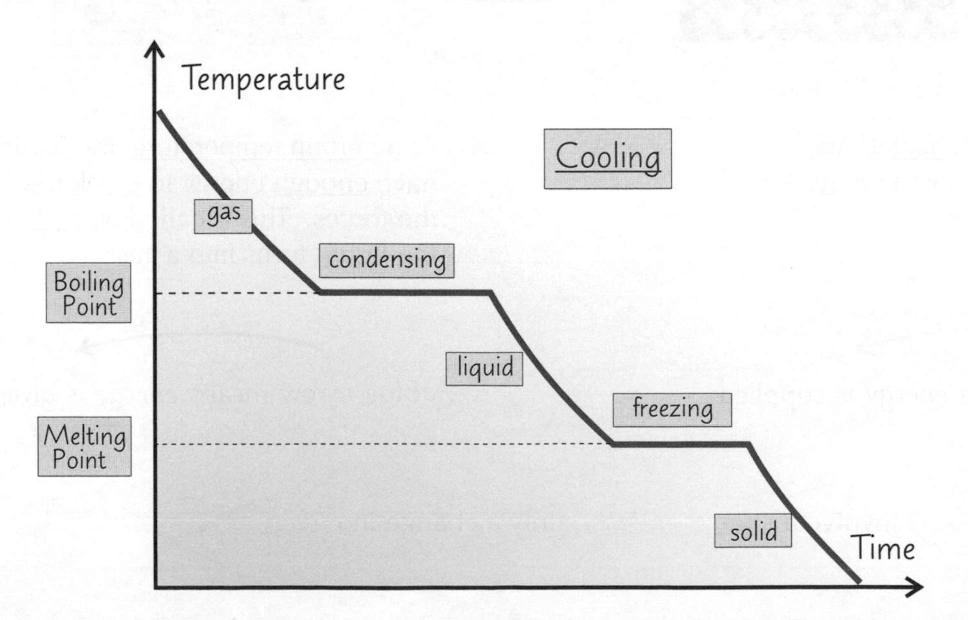

During changes of state, heating and cooling curves go flat

Make sure you understand that when a substance is heated, its temperature increases until it starts melting or boiling. Then the temperature stays the same while the energy is used to weaken forces between particles, giving a flat bit on the curve. The opposite happens when a substance cools.

Warm-Up and Practice Questions

There's a bit too much gas in this section in my opinion. Just tackle the Warm-Up Questions first, then move on to the trickier Practice Questions.

Warm-Up Questions

1) Name the only state of matter that can be easily compressed.
2) In which state of matter are the particles fixed in a regular pattern?
3) What happens to the speed at which particles move when they are heated?
4) What is sublimation?
5) Why does gas pressure increase when the volume is decreased?

Practice Questions

1 An aerosol can of deodorant contains liquefied gas under pressure.

(a) The aerosol can is a solid. Which two of the following properties are properties of a solid?

Has a definite volume | Flows easily

Is easily compressed | Has a high density

Takes the shape of its container

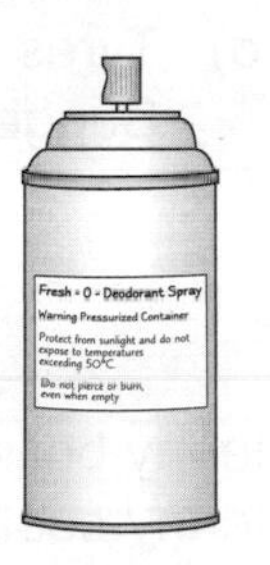

(2 marks)

(b) When the deodorant is sprayed, it changes from a liquid into a gas. Each deodorant particle can be represented by a circle. Copy and complete the diagrams below to show the arrangement of particles in the deodorant, as a liquid and as a gas.

gas | liquid

(2 marks)

(c) A student uses the deodorant in the corner of a changing room. After a while everyone in the room can smell the deodorant.

(i) Name the process that causes gas particles to spread through the room.

(1 mark)

(ii) Describe how the process works.

(2 marks)

Practice Questions

2 Drivers are advised to check their car's tyre pressures before setting out on a long journey, particularly if the car is heavily loaded.

(a) Explain how air molecules inside a car tyre exert pressure on the walls of the tyre.

(1 mark)

(b) In terms of the gas particles, explain why tyre pressure is increased when a car is heavily loaded.

(2 marks)

(c) After several hours driving, the tyres feel hot to the touch.

(i) What effect will this have on the speed of the air molecules inside the tyre?

(1 mark)

(ii) In terms of the motion of the air molecules, give one reason why this will lead to an increase in tyre pressure.

(1 mark)

(d) Tyres at very high pressure can be dangerous. Suggest what might happen to a tyre if the pressure is too high.

(1 mark)

3 Jenny boils 2 litres of water in a large pan in her kitchen. After half an hour Jenny cools the water and measures it in a jug. There is 1 litre left.

(a) Explain what has happened to the water that is not in the jug.

(1 mark)

(b) Jenny notices that there are droplets of water on her kitchen window. Name the process that has taken place to form the droplets.

(1 mark)

(c) Jenny freezes the water in the jug to make some ice cubes.

(i) Do the particles change when the water freezes? Explain your answer.

(1 mark)

(ii) Sketch a cooling curve for the water as it freezes and continues to cool. Include the following labels: liquid, solid, freezing, melting point.

(2 marks)

Atoms and Elements

If you've ever wondered what everything is made of, then the simple answer is atoms.

You Need to Know About **Atoms...**

1) Atoms are a type of tiny, tiny, particle.
2) They're so small that you can't see them directly. So for a long time, no one knew much about them.
3) Dalton was the first modern scientist to try to explain things about atoms.
According to the Dalton model:

- All matter is made up of atoms.
- There are different types of atom.
- Each element (see below) contains a different type.

Scientists now know a lot more about atoms — but luckily, this is all you need to learn for Key Stage 3.

...and **Elements**

1) An element is a substance that contains only one type of atom.
2) Quite a lot of everyday substances are elements:

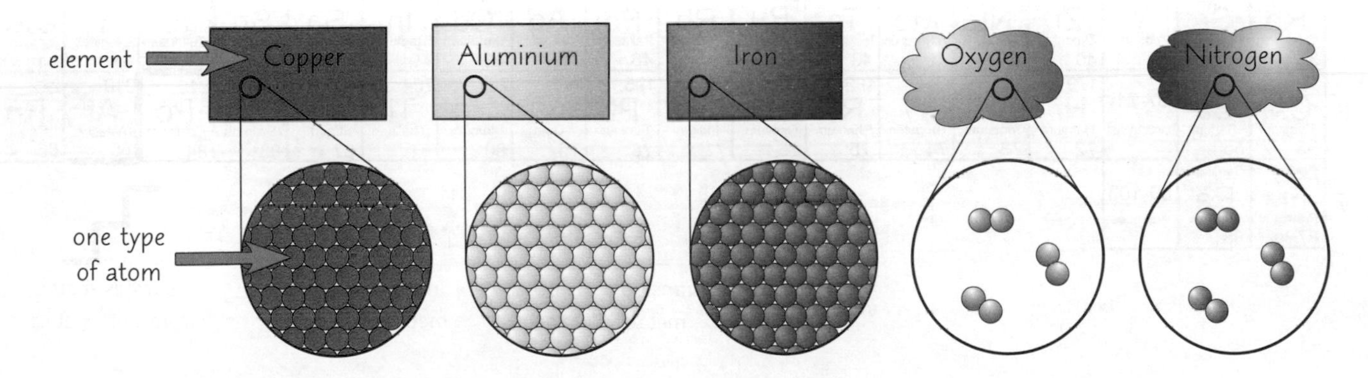

3) All of these elements have different properties.
For example, copper is a soft, bendy metal. Oxygen is a colourless gas.

All Elements Have a **Name** and a **Symbol**

1) There are over 100 different elements and writing their names out each time you wanted to mention one would take ages.
2) So each element has a symbol — usually of one or two letters.

Some symbols make sense (like O for oxygen) but others are based on Latin, so are a bit weird — like Fe for iron.

Examples: Oxygen has the symbol O. Helium has the symbol He.
Carbon has the symbol C. Iron has the symbol Fe.

3) You can see the symbol for each element on the periodic table (see next page).

The Periodic Table

The **Periodic Table** Lists **All** the **Elements**

1) The periodic table shows all the elements we have discovered.
2) The first version of the table was put together by a scientist called Mendeleev. It's thanks to Mendeleev that elements with very similar properties are arranged into vertical columns in the table.
3) The vertical columns are called groups.
4) The horizontal rows are called periods.
5) If you know the properties of one element in a group, you can predict the properties of other elements in that group. E.g. Group 1 elements are all soft, shiny metals, which react in a similar way with water.

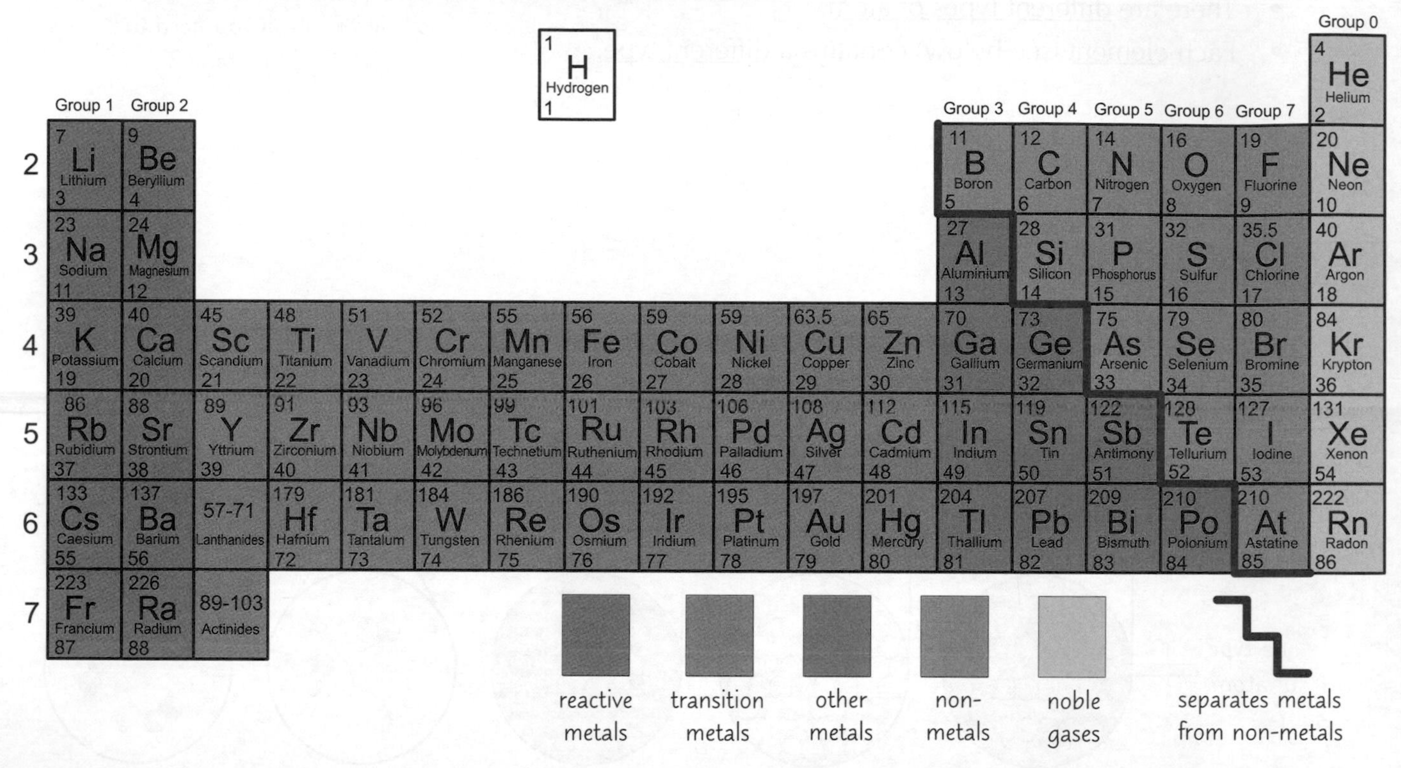

You Can Use the Periodic Table to **Predict Patterns** in **Reactions**

1) In a chemical reaction, elements combine to form new substances (see next page).
2) An element that's dead keen to combine with other elements is said to be very reactive. Group 1, 2 and 7 elements are all pretty reactive.
3) Group 0 elements (the "noble gases") are all extremely unreactive. They almost never take part in any chemical reactions.
4) You can use the periodic table to predict patterns in chemical reactions. For example...

The Group 1 metals get MORE reactive as you go down the group. You can see this by the way the Group 1 metals react with water. When lithium (Li) reacts with water, it fizzes. When rubidium (Rb) reacts with water, it explodes. This is because rubidium is much more reactive than lithium.

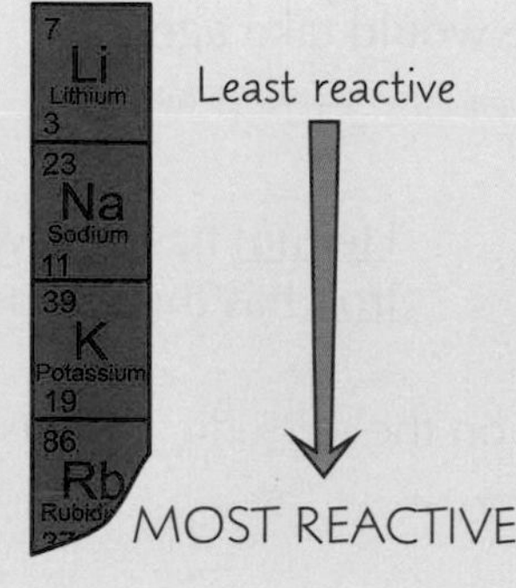

The non-metals in Group 7 behave in the opposite way to the metals in Group 1. They get LESS reactive as you go down the group.

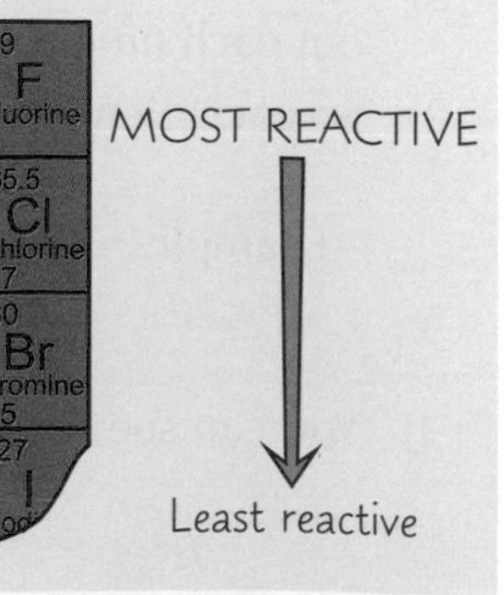

Compounds

Compounds form when different atoms join together.

Compounds Contain Two or More Elements Joined Up

1) When two or more atoms join together, a molecule is made.
   The join is known as a chemical bond.
2) Compounds are formed when atoms from different elements join together. Like in CO_2.

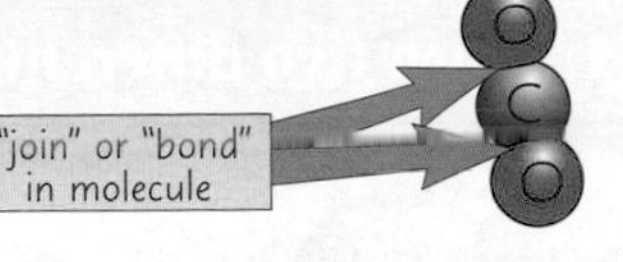

An element which is made up of molecules:

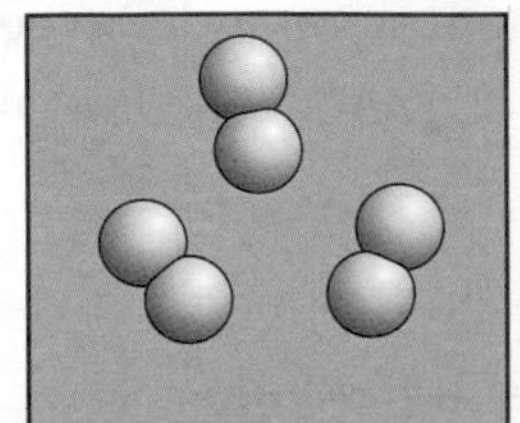

The atoms are joined, but they're all the same, so it's an element.

Molecules in a compound:

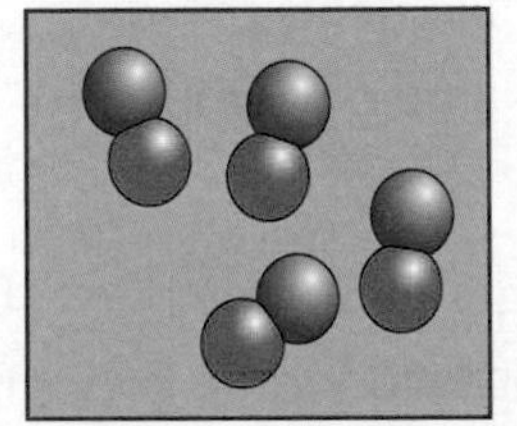

Here we have different atoms joined together — that's a compound alright.

A mixture of different elements:

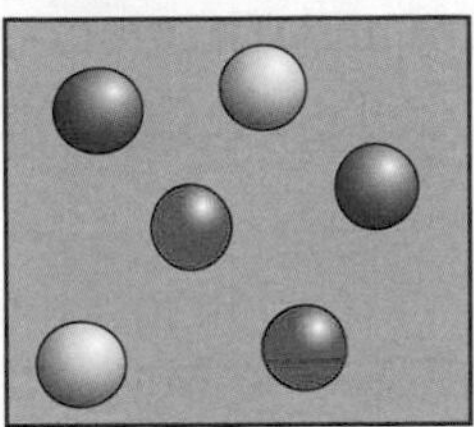

This is not a compound because the elements aren't joined up — it's a mixture (p.60).

Compounds are Formed from Chemical Reactions

1) A chemical reaction involves chemicals (called the reactants) combining together or splitting apart to form one or more new substances (called the products).
2) When a new compound is synthesised (made), elements combine.
3) The new compounds produced by any chemical reaction are always totally different from the original elements (or reactants).
   The classic example of this is iron reacting with sulfur as shown below.

Iron's Properties Change when it Forms a Compound

Iron is magnetic. It reacts with sulfur to make iron sulfide, a totally new substance which is not magnetic.
These equations show what happens in the reaction:

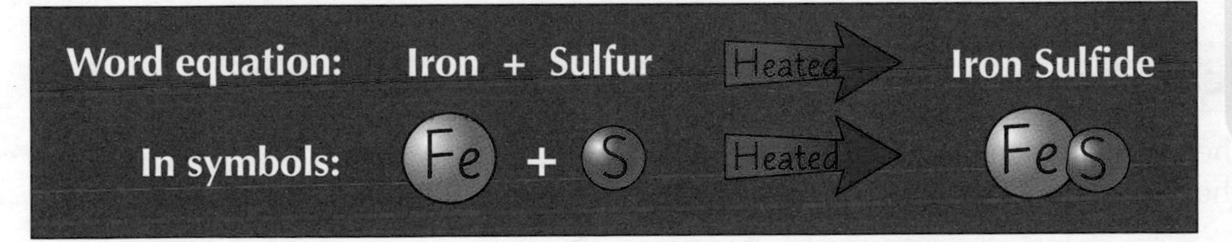

Remember, every element has a name and a symbol. See p.55 for more.

1) When elements undergo a chemical reaction like the one above, the products will always have a chemical formula — e.g. H_2O for water or FeS for iron sulfide.
2) Compounds can be split up back into their original elements but it won't just happen by itself — you have to supply a lot of energy to make the reaction go in reverse.

Chemical reactions form brand new products

Don't get confused between elements, compounds and mixtures. If you're struggling to remember which is which, cover the page and practice writing out what each one means.

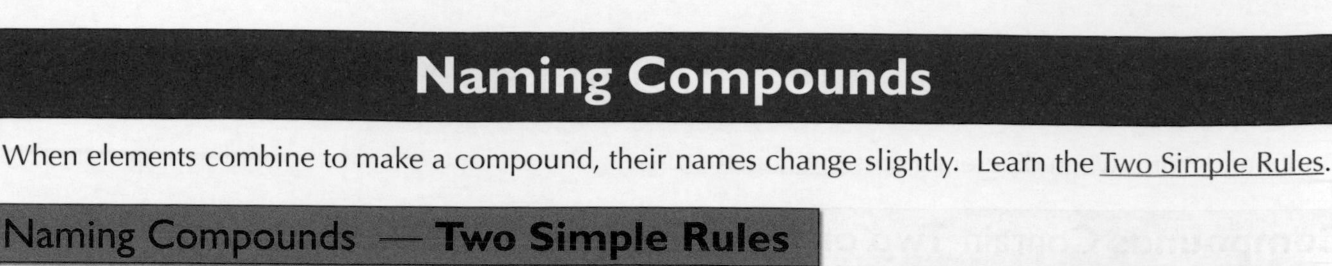

Naming Compounds

When elements combine to make a compound, their names change slightly. Learn the Two Simple Rules.

Naming Compounds — **Two Simple Rules**

Rule 1: When two different elements combine the ending is usually "something -ide".

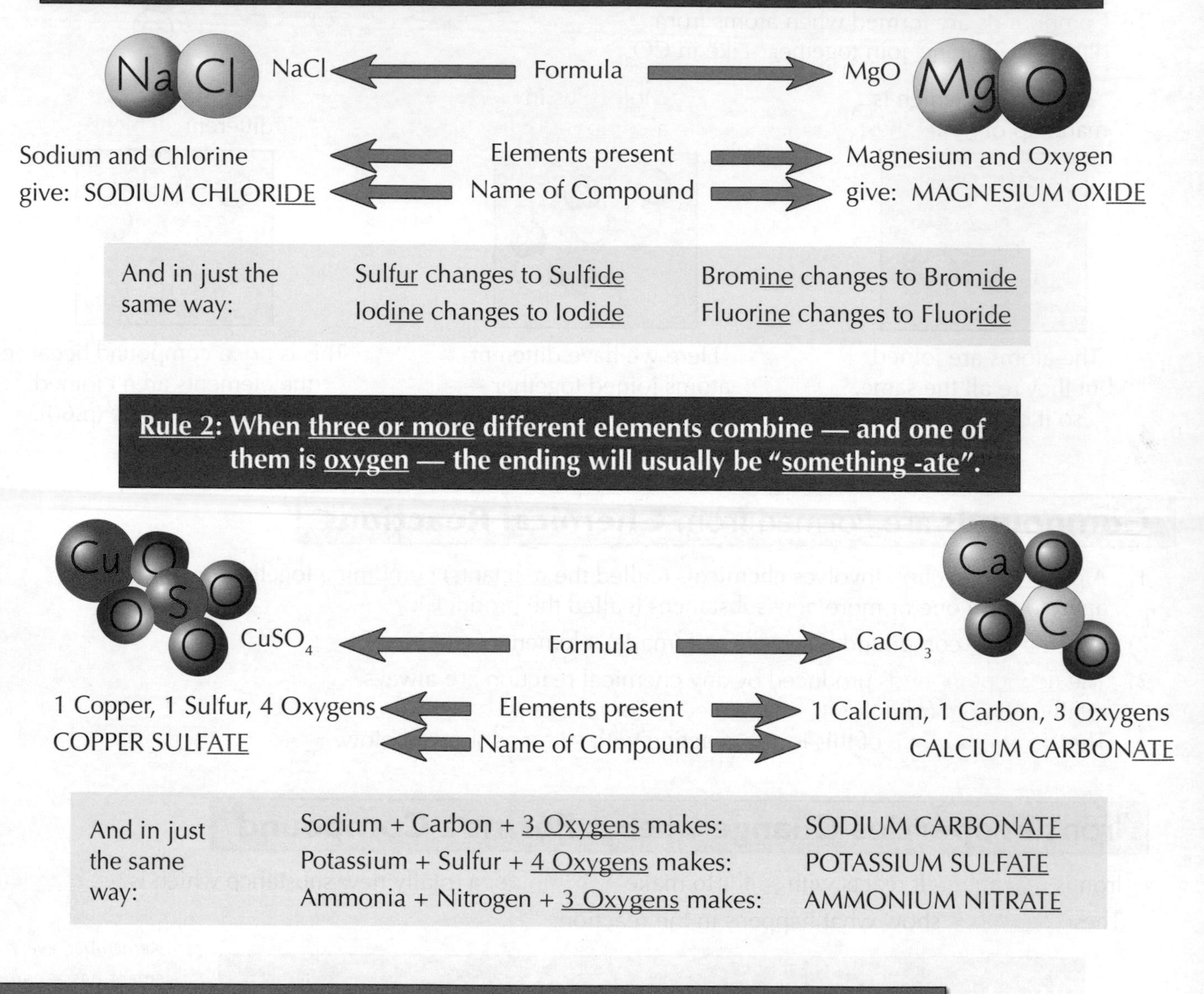

And in just the same way:	Sulfur changes to Sulfide	Bromine changes to Bromide
	Iodine changes to Iodide	Fluorine changes to Fluoride

Rule 2: When three or more different elements combine — and one of them is oxygen — the ending will usually be "something -ate".

And in just the same way:		
	Sodium + Carbon + 3 Oxygens makes:	SODIUM CARBONATE
	Potassium + Sulfur + 4 Oxygens makes:	POTASSIUM SULFATE
	Ammonia + Nitrogen + 3 Oxygens makes:	AMMONIUM NITRATE

If **Two Identical** Elements **Combine**, it's Not a **Compound**

Identical atoms of the same element are often found combined.
This doesn't make them a compound though — in fact, their name doesn't even change.

H_2 = Hydrogen (H H)
N_2 = Nitrogen (N N)
O_2 = Oxygen (O O)
F_2 = Fluorine (F F)
Cl_2 = Chlorine (Cl Cl)
Br_2 = Bromine (Br Br)

These are all elements with two atoms, not compounds. They're almost never found as single atoms in nature.

WORKING SCIENTIFICALLY

What's in a name?

Turns out, lots. It's important to be able to name chemicals correctly, so learn the rules.

Warm-Up and Practice Questions

Here are five simple questions to get you going, then a couple of more challenging Practice Questions to make sure you really understand atoms, elements, compounds and the periodic table.

Warm-Up Questions

1) What is an element?
2) Is bromine more or less reactive than fluorine?
3) Which group of the periodic table contains only extremely unreactive elements?
4) What is the difference between a compound and a mixture?
5) What is the name of a compound made up of iron, sulfur and oxygen?

Practice Questions

1 The diagrams below represent the arrangement of atoms and molecules in four different substances, A, B, C and D.

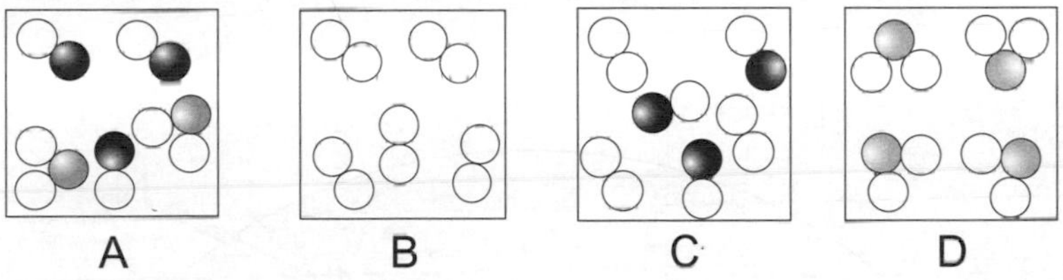

A B C D

(a) Which substance is a pure element?

Substance B

(1 mark)

(b) Which substance is a mixture of compounds?

Substance A

(1 mark)

(c) Which substance would you expect to find in the periodic table?

Substance B

(1 mark)

(d) Which substance is most likely to be water, H_2O?

Substance D

(1 mark)

2 Copper is a pinky-orange coloured metal.
Compounds of copper can form coloured crystals.

(a) When copper is heated in oxygen, layers of a black substance form on the metal. Suggest what the black substance that forms during the reaction is.

Copper Oxide

(1 mark)

(b) Suggest what substance is formed when copper is reacted with carbon and oxygen.

Copper carbonate

(1 mark)

Mixtures

Mixtures in chemistry are like cake mix in the kitchen — all the components are mushed up together, but you can still pick out the raisins if you really want. You'll need to learn the technical terms too though...

Mixtures are Substances That are **NOT** Chemically **Joined Up**

1) A pure substance is made up of only one type of element OR only one type of compound. It can't be separated into anything simpler without a chemical reaction.

E.g. pure water is made up of H_2O molecules only. These molecules can't be separated into H and O atoms without a chemical reaction.

2) A mixture contains two or more different substances. These substances aren't chemically joined up — so, if you're clever, you can separate them very easily using physical methods (i.e. without a chemical reaction). See pages 62-64 for more.
3) Sea water and air are good examples of mixtures — they contain several different substances which aren't chemically combined.
4) A mixture has the properties of its constituent parts (i.e. the parts it's made from).

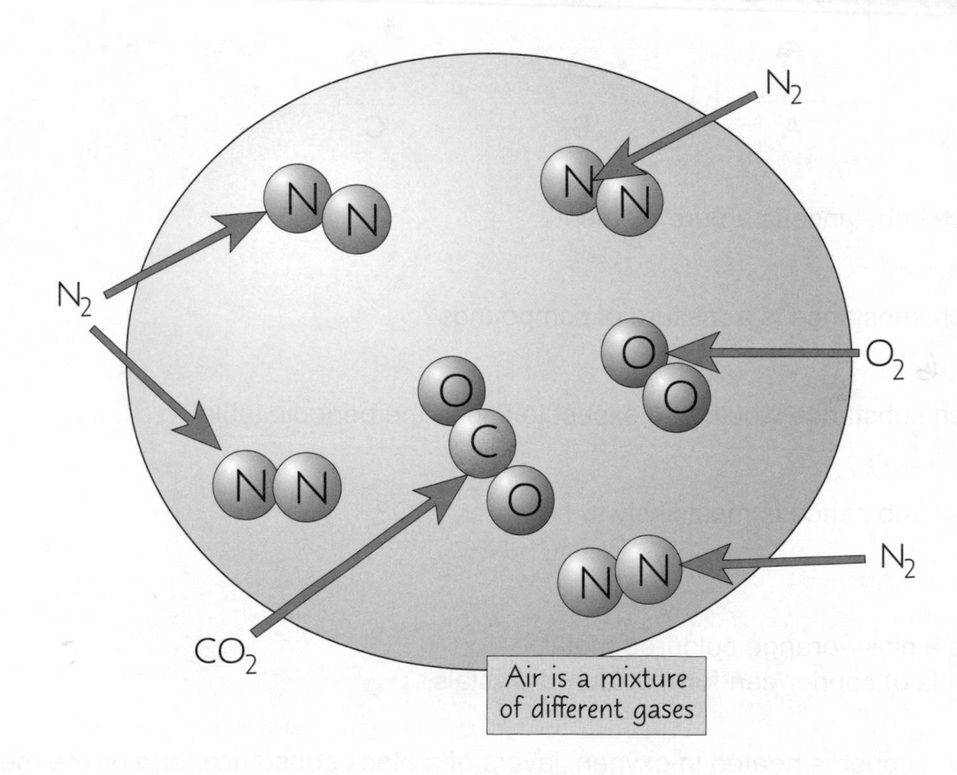

Air is a mixture of different gases

The components of a mixture are not chemically combined

I've said it already, but this is important — the parts of a mixture are not chemically joined up at all. You can separate the substances relatively easily using physical methods (more on them later). This makes them very different to compounds, which need a chemical reaction to separate them.

Mixtures

Dissolving isn't Disappearing

1) Dissolving is a common way mixtures are made.
2) When you add a solid (the solute) to a liquid (the solvent) the bonds holding the solute particles together sometimes break.
3) The solute particles then mix with the particles in the liquid, forming a solution.

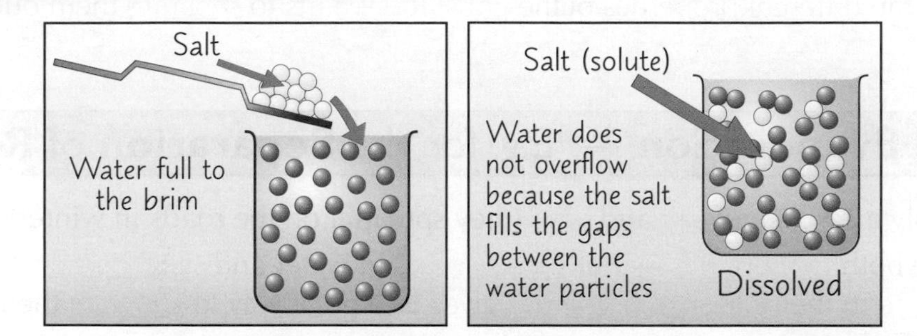

Learn these seven definitions:

1) Solute – is the solid being dissolved.
2) Solvent – is the liquid it's dissolving into.
3) Solution – is a mixture of a solute and a solvent that does not separate out.
4) Soluble – means it WILL dissolve.
5) Insoluble – means it will NOT dissolve.
6) Saturated – a solution that won't dissolve any more solute at that temperature.
7) Solubility – a measure of how much solute will dissolve.

4) Remember, when salt dissolves it hasn't vanished — it's still there — no mass is lost.
5) If you evaporated off the solvent (the water), you'd see the solute (the salt) again.

Solubility Increases with Temperature

1) At higher temperatures more solute will dissolve in the solvent because particles move faster.
2) However some solutes won't dissolve in certain solvents. E.g. salt won't dissolve in petrol.

There is no change in mass when a solid dissolves

It might look like salt disappears in water, but it's still there and it still has a mass. A given amount of water can only dissolve a certain amount of salt — but you can increase this amount by heating the water. Make sure you remember that, and learn the seven terms in the box.

Separating Mixtures

There are all sorts of ways you can separate mixtures. You've got to know four of them.

Mixtures Can be Separated Using Physical Methods

There are four separation techniques you need to be familiar with.

1) FILTRATION 2) EVAPORATION 3) CHROMATOGRAPHY (page 63) 4) DISTILLATION (page 64)

All four make use of the different properties of the constituent parts to separate them out.

Filtration and Evaporation — E.g. for the Separation of Rock Salt

1) Rock Salt is simply a mixture of salt and sand (they spread it on the roads in winter).
2) Salt and sand are both compounds — but salt dissolves in water and sand doesn't. This vital difference in their physical properties gives us a great way to separate them.

You Need to Learn the Four Steps of the Method:

1) Grinding 2) Dissolving 3) Filtering 4) Evaporating

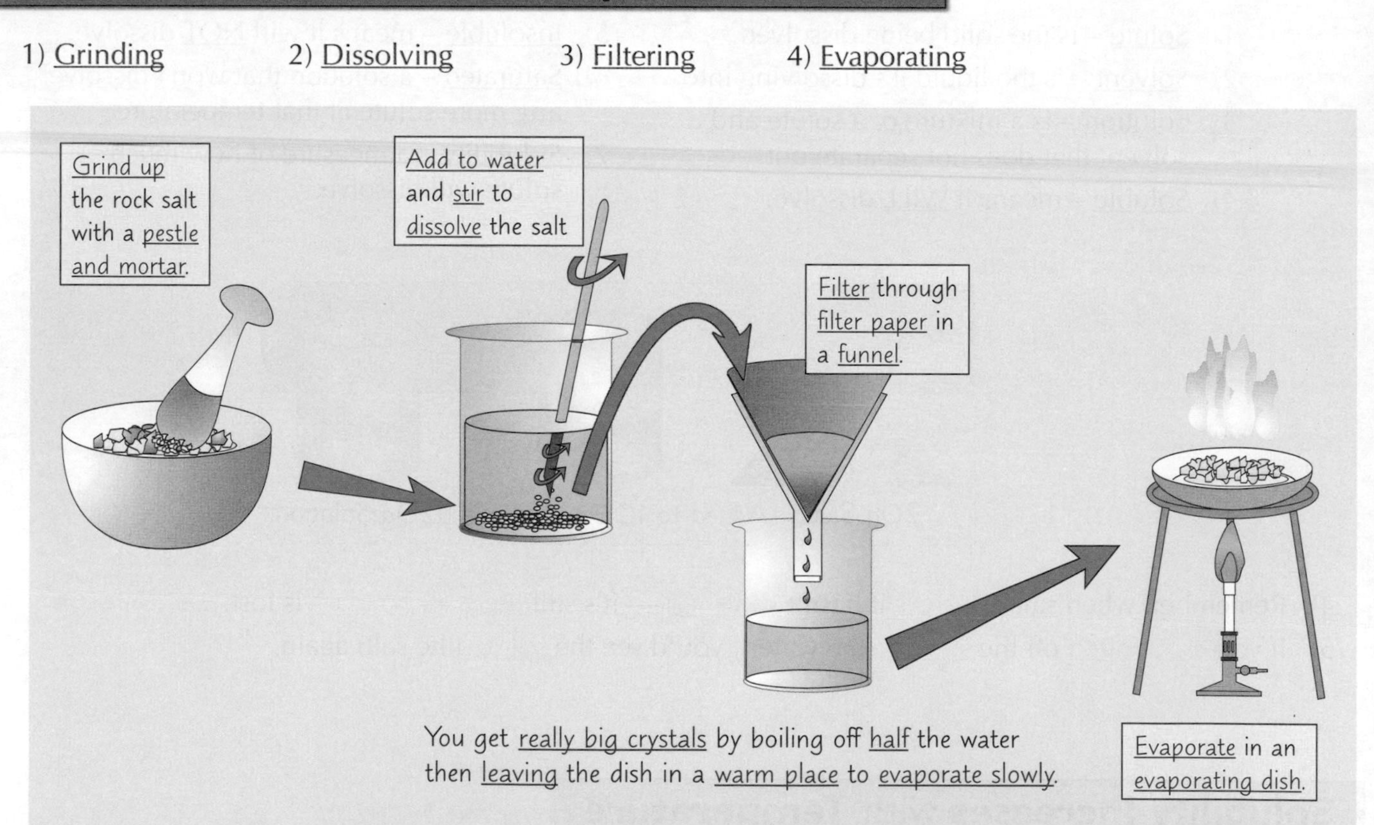

- The sand doesn't dissolve (it's insoluble) so it stays as big grains and obviously these won't fit through the tiny holes in the filter paper — so it collects on the filter paper.
- The salt is dissolved in solution so it does go through — and when the water's evaporated, the salt forms as crystals in the evaporating dish. This is called crystallisation. (Surprise surprise.)

Grind, dissolve, filter, evaporate.

It's pretty easy to separate rock salt into rock (sand) and salt. Salt dissolves in water, but sand does not. So all you need to do is mash up the rock salt, dissolve the salt, fish out the sand with a filter and then get rid of the water by evaporating it off. Easy when you know how — make sure you do.

Separating Mixtures

Chromatography is all about separating different liquids, like ink dyes.

Chromatography is Ideal for Separating Dyes in Inks

1) Different dyes in ink will wash through paper at different rates.
2) Some will stick to the paper and others will remain dissolved in the solvent (see below) and travel through it quickly.

Method 1

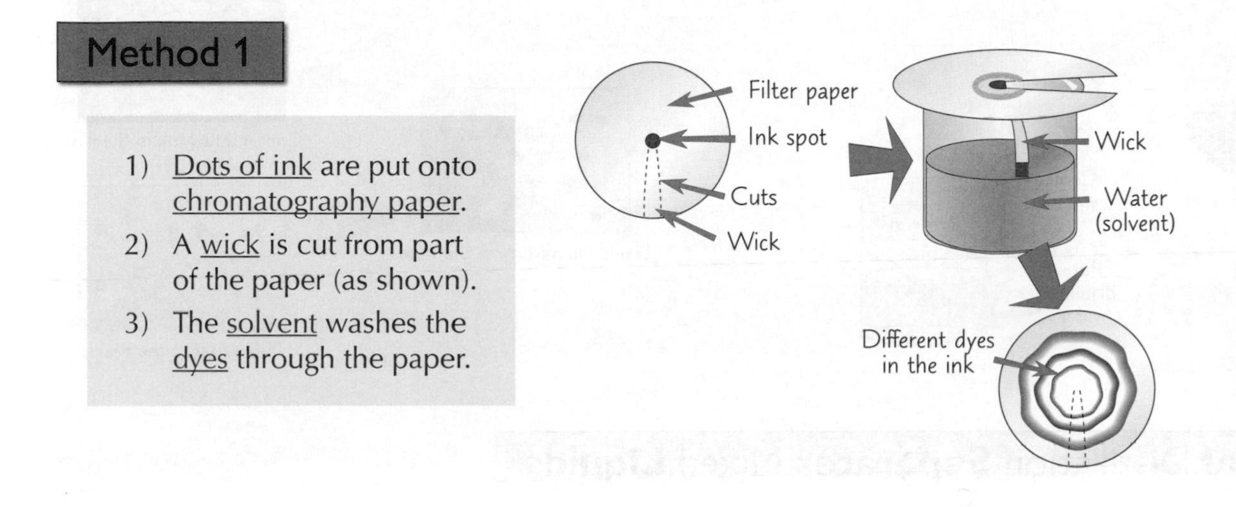

1) Dots of ink are put onto chromatography paper.
2) A wick is cut from part of the paper (as shown).
3) The solvent washes the dyes through the paper.

Method 2

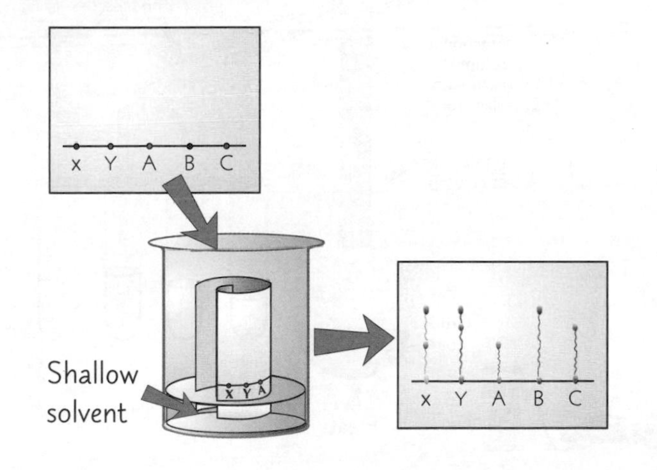

1) Put spots of inks onto a pencil baseline on chromatography paper.
2) Roll the sheet up and put it in a beaker.
3) The solvent seeps up the paper, carrying the ink dyes with it.
4) Each different dye will form a spot in a different place.
5) You can compare a forged ink to a known ink to see which it is.

Two other exciting uses of chromatography are:

1) **Identifying blood samples.**
2) **Investigating chlorophyll.**

Remember — chromatography is for separating liquids

It's important to think about why you do each stage of an experimental method. For example, in method 2 the baseline is drawn in pencil. If it were drawn in pen it might be carried up the paper with the ink spots and make the separation of the spots hard to see.

Separating Mixtures

Simple Distillation **Separates Pure Water** from **Ink**

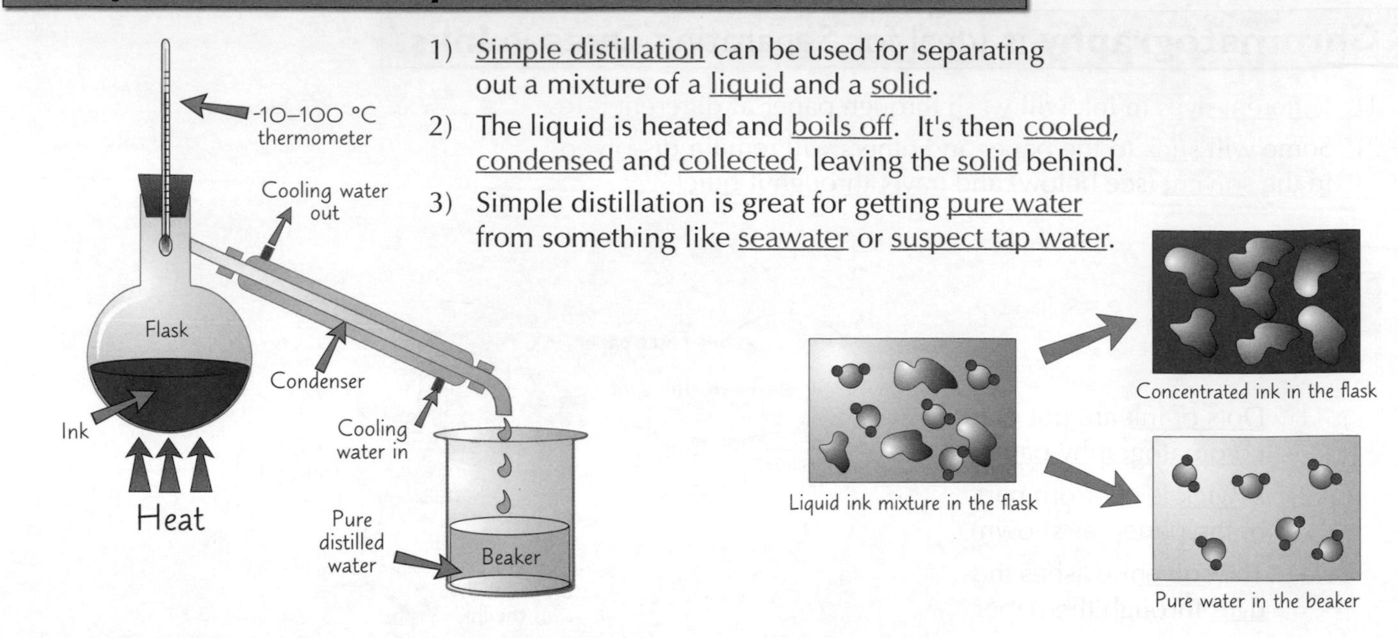

1) Simple distillation can be used for separating out a mixture of a liquid and a solid.
2) The liquid is heated and boils off. It's then cooled, condensed and collected, leaving the solid behind.
3) Simple distillation is great for getting pure water from something like seawater or suspect tap water.

Fractional Distillation **Separates** Mixed **Liquids**

1) Fractional distillation is used for separating a mixture of liquids like crude oil.
2) Different liquids will boil off at different temperatures, around their own boiling point.
3) The fractionating column ensures that the "wrong" liquids condense back down, and only the liquid properly boiling at the temperature on the thermometer will make it to the top.
4) When each liquid has boiled off, the temperature reading rises until the next fraction starts to boil off.
5) Real-life examples include:
   - distilling whisky,
   - separating crude oil into petrol, diesel and other fuels.

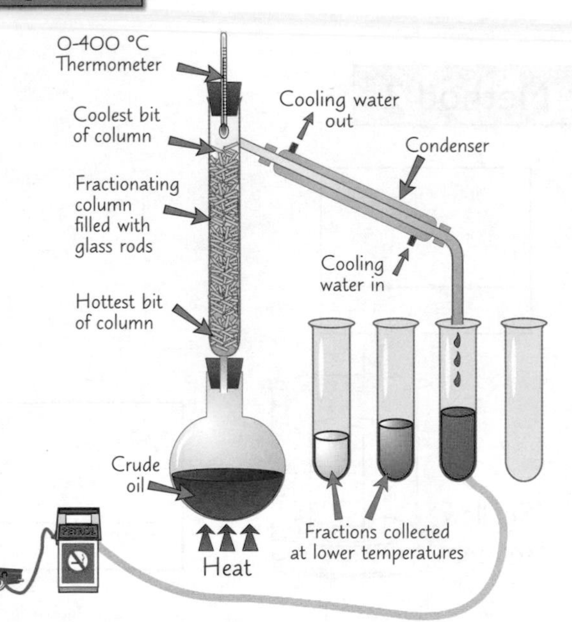

Check **Purity** with **Melting** and **Boiling Points**

1) A pure chemical substance has fixed melting and boiling points. E.g. pure water boils at 100 °C and pure ice melts at 0 °C. These figures are known for a huge range of substances.
2) This helps us to identify unknown substances, e.g. if a liquid boils at exactly 100 °C it's likely to be pure water.
3) Impurities change melting and boiling points, e.g. impurities in water cause it to boil above 100 °C.
4) This means you can test the purity of a substance you've separated from a mixture.

Pure Substance	Melting Point °C	Boiling Point °C
Water	0	100
Ethanol	-114	78
Aluminium	660	2520

Warm-Up and Practice Questions

Five pages later, it's time for some more questions to check you're learning as well as reading. If you don't know the answer to any of the questions, go back and read the pages again.

Warm-Up Questions

1) Why are mixtures (usually) easier to separate than compounds?
2) What is a solute? And what is a solvent?
3) What happens to solubility when the temperature decreases?
4) Describe the four steps you would take to obtain salt from rock salt.
5) What is chromatography used to do?
6) Which separation technique is used to separate crude oil?

Practice Questions

1 Amanda mixes excess copper oxide with dilute sulfuric acid until no more copper oxide will dissolve. She is left with a blue solution of copper sulfate, mixed with unreacted copper oxide powder. She separates the copper oxide powder from the copper sulfate solution by using the apparatus shown below:

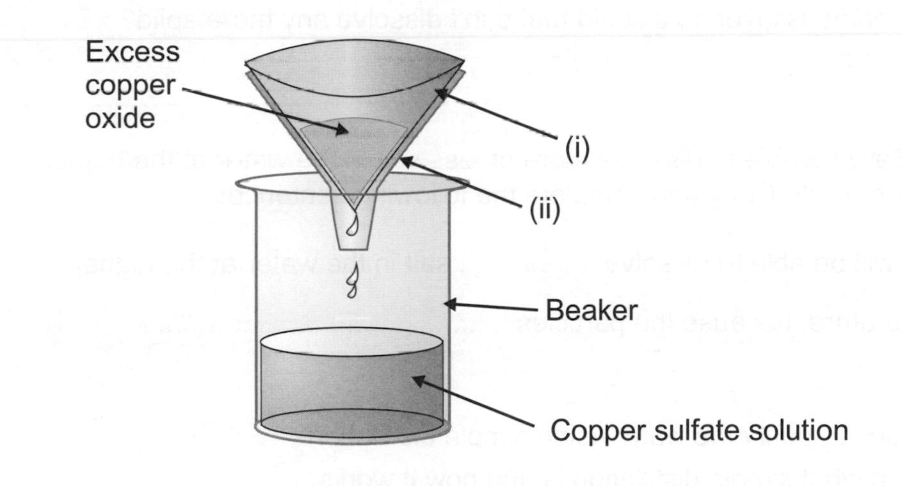

(a) Write down what the missing labels on the diagram above should be.

(2 marks)

(b) What is this separation technique called?

(1 mark)

(c) Amanda wants to obtain large copper sulfate crystals from the solution. Suggest how she could do this with a simple experimental technique.

(1 mark)

Practice Questions

2 When Sally adds salt to water and stirs it, the salt dissolves in the water.

(a) Copy and complete the table below:

Scientific term	Substance
solute	
solvent	
	salt water

(3 marks)

(b) Sally measures the maximum amount of salt that can be dissolved in 100 ml of water at room temperature. She then repeats her experiment with 100 ml of water heated to 60 °C.

(i) What name is given to a liquid that can't dissolve any more solid?

(1 mark)

(ii) Will Sally be able to dissolve more or less salt in the water at the higher temperature? Copy and complete the following sentence:

Sally will be able to dissolve .... salt in the water at the higher temperature, because the particles ....

(2 marks)

(c) Sally decides to purify the water using simple distillation.

(i) Explain what simple distillation is and how it works.

(3 marks)

(ii) Suggest how Sally could check that the distilled water is pure.

(1 mark)

Properties of Metals

1) **Metals** Can be Found in the **Periodic Table**

1) Most of the elements in the periodic table are metals.
2) Some are shown here in blue, to the left of the zig zag.

Li	Be														
Na	Mg											Al			
K	Ca	Sc	Ti	V	Cr	Mn	Fe	Co	Ni	Cu	Zn	Ga	Ge		
Rb	Sr	Y	Zr	Nb	Mo	Tc	Ru	Rh	Pd	Ag	Cd	In	Sn	Sb	
Cs	Ba	La	Hf	Ta	W	Re	Os	Ir	Pt	Au	Hg	Tl	Pb	Bi	Po
Fr	Ra	Ac													

2) **Metals** Conduct **Electricity**

1) Electric current is the flow of electrical charge around a circuit.
2) Metals all allow electrical charge to pass through them easily.
3) The moving charges are negatively-charged particles called electrons.
4) Metals contain some electrons that are free to move between the metal atoms. These free electrons can carry an electric current from one end of the metal to the other.
5) Because they conduct electricity well, metals are often used to make wires and parts of electrical circuits. Copper is a good example of a metal used in this way.

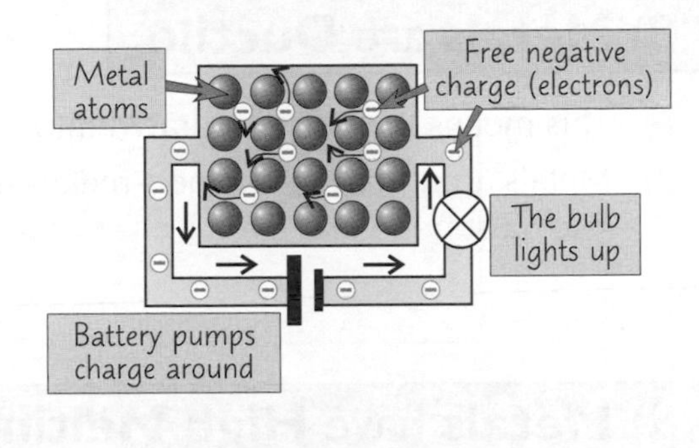

3) **Metals** Conduct **Energy**

1) Metals transfer energy from a hot place to a cold place quickly and easily.
2) The "hot" particles vibrate strongly.
3) Because the particles are very close together, the vibrations are easily passed on through the metal.
4) Free electrons in the metal also help to transfer energy from the hot parts of the metal to the cooler parts as they move around.
5) This is why you get things like saucepans made out of metal, e.g. aluminium. Energy from the hob passes easily through the pan and cooks your spaghetti.

4) **Metals** are **Strong** and **Tough**

1) Metals have high tensile strength (they can be pulled hard without breaking).
2) This is because there are strong forces between metal atoms that hold them together.
3) So they make good building materials.

5) **Metals** are **Shiny** When **Polished**

Polished or freshly cut metals give strong reflection of light from their smooth surface. This makes them look shiny.

6) **Metals** are **Sonorous**

This means they make a nice "donnnnggg" sound when they're hit. If you think about it, it's only metals that do that — you could make a gong out of plastic, but it wouldn't be much good.

Properties of Metals

7) Metals are Malleable

1) Metals are easily shaped (malleable) because the atoms in metals can slide over each other.
2) This means metals can be hammered into thin sheets or bent — all without shattering.

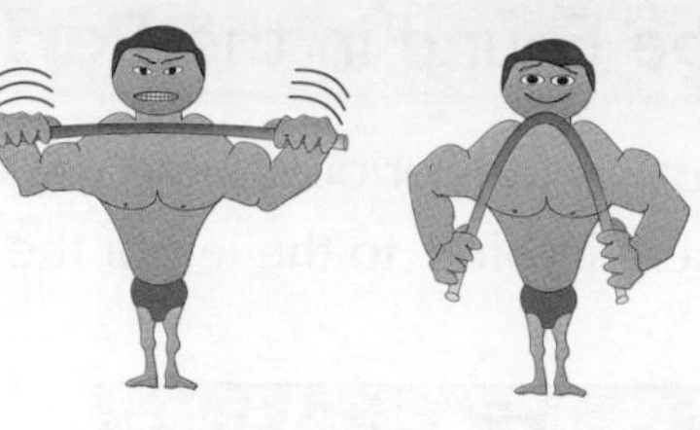
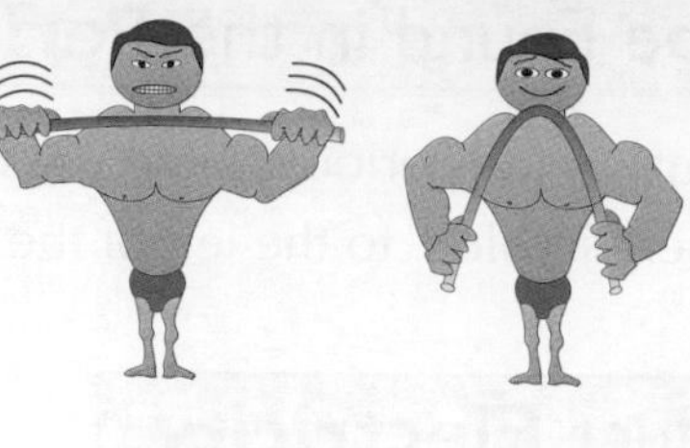

8) Metals are Ductile

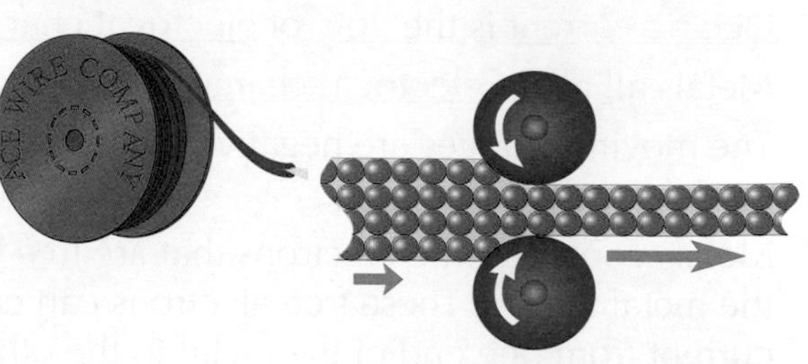

1) This means they can be drawn into wires.
2) Metals aren't brittle like non-metals (see page 70) are. They just bend and stretch.

9) Metals have High Melting and Boiling Points

1) A lot of energy is needed to melt metals.
2) This is because their atoms are joined up with strong bonds.
3) The table shows how hot they have to get to melt.

Metal	Melting Point (°C)	Boiling Point (°C)
Aluminium	660	2520
Copper	1085	2562
Magnesium	650	1090
Iron	1538	2861
Zinc	420	907
Silver	962	2162

10) Metals have High Densities

1) Density is all to do with how much stuff there is squeezed into a certain space.
2) Metals feel heavy for their size (i.e. they're very dense) because they have a lot of atoms tightly packed into a small volume.

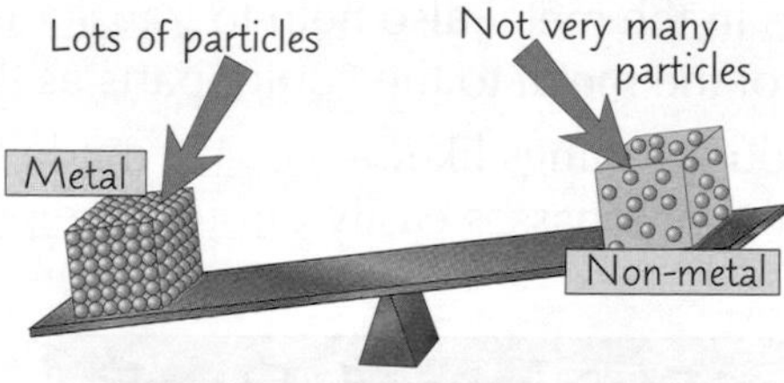

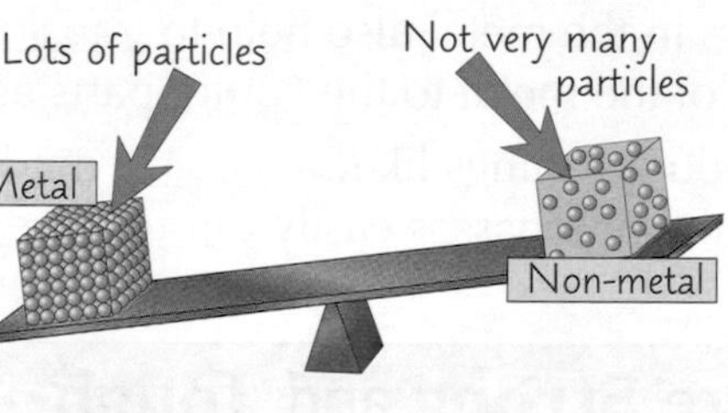

11) Metals Make Alloys When Mixed with Other Metals

1) A combination of different metals is called an alloy. The properties of the metals get jumbled up in the new alloy.
2) So lighter, weaker metals can be mixed with heavier, stronger metals and the result is, hopefully, an alloy which is light and strong.

Alloy Wheels — light and strong

12) Some Metals are Magnetic

1) Only certain metals are magnetic.
2) Most metals aren't magnetic. Iron, nickel and cobalt are. Alloys made with these three metals will also be magnetic — e.g. steel is made mostly from iron, so is also magnetic.

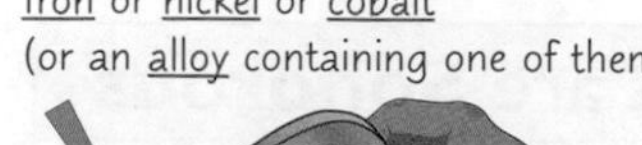

Properties of Non-metals

The properties of non-metal elements vary quite a lot. As you will slowly begin to realise...

1) **Non-metals** Can be Found in the **Periodic Table**

1) All the non-metals (with the exception of hydrogen) are clustered in the corner over on the right of the zigzag. Look, right over there.
2) There are fewer non-metals than metals.

2) **Non-metals** are **Poor Conductors** of **Electricity**

1) Most non-metals are insulators, which means that charges can't flow through them.
2) If charges can't move then no electric current flows. This is very useful — non-metals combine to make things like plugs and electric cable coverings.

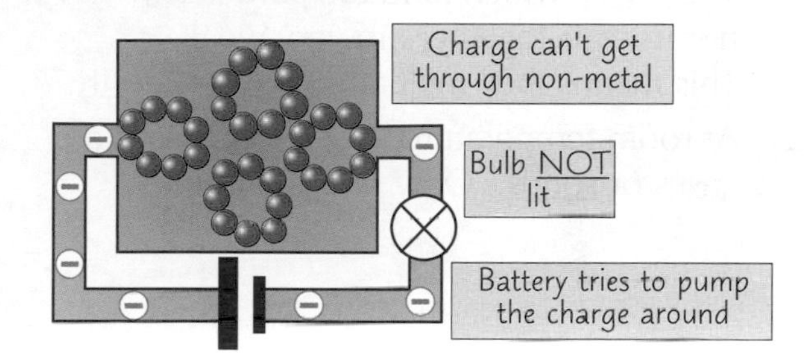

One exception to this rule is graphite — a non-metal made purely from carbon atoms. Its atoms are arranged in layers, which allow electrons to move along them, so graphite can conduct electricity.

3) **Non-metals** are **Poor Conductors** of **Energy** by **Heating**

1) Non-metals don't transfer energy from a hot place to a cold place very quickly or easily.
2) This makes non-metals really good insulators.
3) "Hot" particles don't pass on their vibrations so well.

4) **Non-metals** are **not Strong** or **Hard-Wearing**

1) The forces between the particles in most non-metals are weak — this means they break easily.
2) It's also easy to scrub atoms or molecules off them — so they wear away quickly.

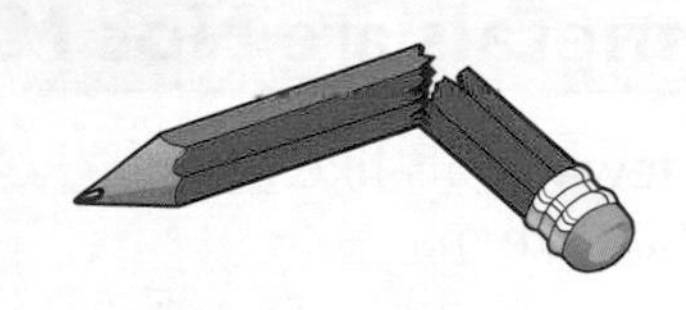

5) **Non-metals** are **Dull**

1) Most non-metals don't reflect light very well at all. Their surfaces are not usually as smooth as metals.
2) This makes them look dull.

Properties of Non-metals

6) Non-metals are Brittle

1) Non-metal structures are held together by weak forces.
2) This means they can shatter all too easily.

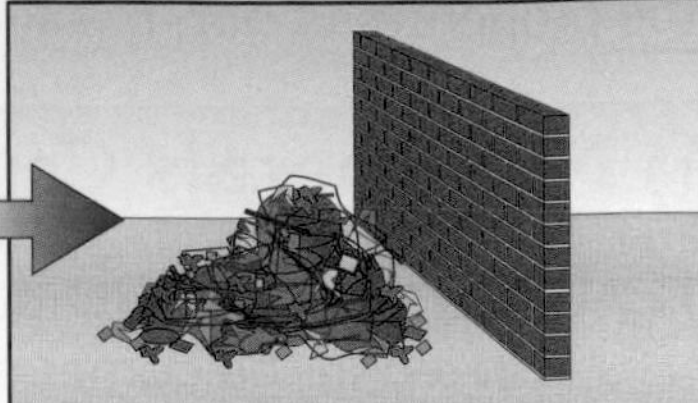

7) Non-metals Have Low Melting Points and Boiling Points

1) The forces which hold the particles in non-metals together are very weak. This means they melt and boil very easily.
2) At room temperature, most non-metals are gases or solids. Only one is a liquid.

Non-Metal	Melting Point (°C)	Boiling Point (°C)
Sulfur	113	445
Oxygen	-218	-183
Chlorine	-101	-35
Helium	-272	-269
Neon	-249	-246
Bromine	-7	59

8) Non-metals Have Low Densities

1) Obviously the non-metals which are gases will have very low density. This means they don't have very many particles packed into a certain space.
2) Some of these gases will even float in air — ideal for party balloons.
3) Even the liquid and solid non-metals have low densities.

9) Non-metals are Not Magnetic

1) Only a few metals like iron, nickel and cobalt are magnetic.
2) All non-metals are most definitely non-magnetic.

Non-metals — nine properties you need to know

There are nine fascinating facts about non-metals to learn here. Why not make a poster with all 9 on, and hang it up where you can see it to help you learn all of them? As a matter of fact, make two posters — one for metals as well. You can never have too many posters.

Properties of Other Materials

As well as metals and non-metals, you need to learn all about some ace compounds and mixtures of compounds — polymers (plastics), ceramics (like bone china) and composites (like fibreglass).

Polymers Have Many Useful Properties

Polymers (that's plastics to you and me) include nylon, polythene and PVC.

1) Polymers are usually insulators — it's difficult for energy to be transferred through them electrically or by heating.
2) They're often flexible — they can be bent without breaking.
3) They have a low density — they can be very light for their size and strength. This makes them ideal for making things that need to be strong but not heavy.
4) They're easily moulded — they can be used to manufacture equipment with almost any shape.

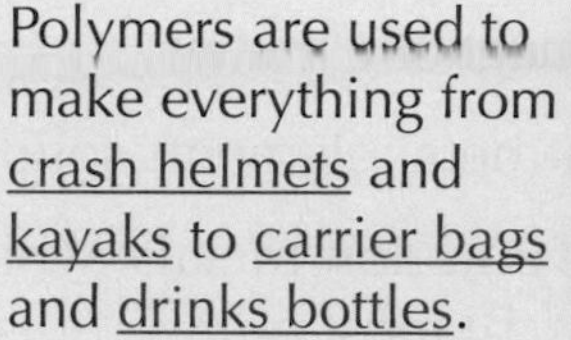

Polymers are used to make everything from crash helmets and kayaks to carrier bags and drinks bottles.

Polymers are just compounds, made by joining loads of little molecules together in long chains. They usually contain carbon.

Ceramics are Stiff but Brittle

Ceramics include glass, porcelain and bone china (for posh tea cups). They are:

1) Insulators — it's difficult for energy to be transferred through them electrically or by heating.
2) Brittle — they aren't very flexible and will break instead of bending.
3) Stiff — they can withstand strong forces before they break.

Ceramics are made by 'baking' substances like clay.

As well as tea cups, ceramics are used for brakes and parts of spark plugs in cars.

Composites are Made of Different Materials

1) Composite materials are made from two or more materials stuck together.
2) This can make a material with more useful properties than either material alone. For example:

Fibreglass

1) Fibreglass (or Glass Reinforced Plastic — GRP) consists of glass fibres embedded in plastic.
2) It has a low density (like plastic) but is very strong (like glass).
3) These properties mean fibreglass is used for things like skis, boats and surfboards.

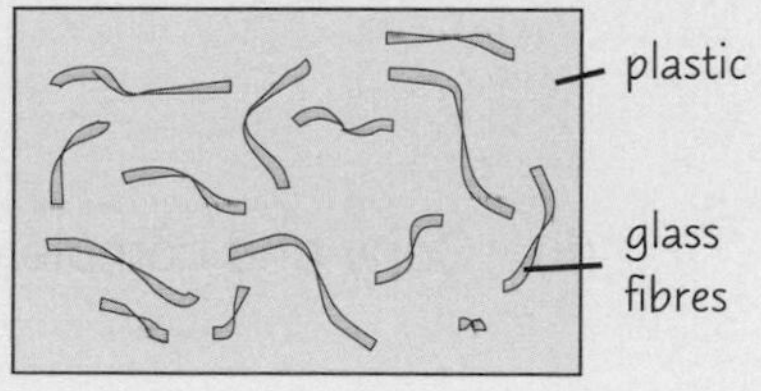

Concrete

1) Concrete is made from a mixture of sand and gravel embedded in cement.
2) It can withstand high compression stresses (i.e. being squashed) so it's great at supporting heavy things. This makes it ideal for use as a building material, e.g. in skate parks, shopping centres, airports, etc.

Warm-Up and Practice Questions

Take a look at these Warm-Up Questions and test yourself. If you get any wrong, go back and have a look at the parts you didn't know, then test yourself again. Simple.

Warm-Up Questions

1) Name three magnetic metals.
2) Are most non-metal elements solids, liquids or gases at room temperature?
3) Out of metals and non-metals, which are the:
   a) densest, b) most brittle, c) shiniest, d) best insulators?
4) Name four useful properties of polymers.
5) What is concrete made from? Why is it a good material for use in buildings?

Practice Questions

1 Metals and non-metals have different properties.

(a) Metals have a number of useful properties.
Give the main reason why:

(i) a bridge is made from metal.

(1 mark)

(ii) a trophy is made from metal.

(1 mark)

(iii) a frying pan is made from metal.

(1 mark)

(b) Sulfur is a non-metal.

(i) Would you expect sulfur to be a good conductor of energy by heating? Explain your answer.

(1 mark)

(ii) Copy and complete the following sentences using words from the box:

shiny	left	brittle
dull	right	flexible

Sulfur can be found on the ....-hand side of the periodic table.

It has a .... consistency and a .... yellow surface.

(3 marks)

Practice Questions

2 Elements can be metals or non-metals.
All known elements are arranged in the periodic table.

He
C
Na
Fe Cu

(a) From the table above, give the symbol of one element which
  (i) is a metal.
  (ii) is a non-metal.
  (iii) is a gas at room temperature.
  (iv) will be attracted by a magnet.

(4 marks)

(b) Does the periodic table contain more metals or non-metals?

(1 mark)

3 Sophie is designing a new bottle for a manufacturer of soft drinks.
(a) The bottle must be light, flexible and have a unique shape.
  (i) Suggest a material that the bottle could be made from.

(1 mark)

  (ii) Give a reason why a metal would not be a good material to use for the bottle.

(1 mark)

(b) The manufacturer also asks Sophie to design a set of cups to mark the release of its new line of hot beverages. It wants the cups to be made of ceramics.
  (i) Suggest why ceramics are a good choice of material for the cups.

(1 mark)

  (ii) Give two other applications in which ceramics are used.

(2 marks)

(c) Sophie is given a surfboard made of fibreglass as a present from the manufacturer when the first bottles and cups are made.
  (i) Fibreglass is a composite material.
  Explain what is meant by a composite material.

(1 mark)

  (ii) Explain why fibreglass is a good choice of material for making surfboards.

(2 marks)

Revision Summary for Section Five

We've moved on to Chemistry now. Makes a refreshing change from all that slimy Biology anyway. Section Five is all about Classifying Materials so here's a whole page of delicious Section Summary questions to help you classify how much you've remembered.

You know the drill: work through these questions and try to answer them. For any you can't do, look back through the section and find the answer — and then learn it. Then try all the questions again and see how many more you can do. Keep going and before you know it you'll be answering them all perfectly.

1) What are the three states of matter? Describe five properties for each of them.
2) Draw a diagram to show how particles are arranged in a solid and in a liquid.
3) Explain how gases exert a pressure on the insides of a container.
4) What happens to the pressure of a gas if the temperature of the gas is increased?
5) Give the names of five changes of state, and say which state they go from and to.
6) For any given substance, in which state do the particles have the most energy?
7) Does a change of state involve a change in mass?
8) Explain why a heating curve has a flat bit when a substance is boiling.
9) What is an atom?
10) Roughly how many elements are there in the periodic table?
11) In the periodic table: a) What is a group? b) What is a period?
12) Using the periodic table, give the chemical symbol for these:
    a) sodium b) magnesium c) oxygen d) iron e) sulfur
    f) aluminium g) carbon h) chlorine i) calcium j) zinc.
13) Which will show a more vigorous reaction when dropped in water, lithium or rubidium? Why?
14) Use the periodic table to predict whether fluorine or iodine is more reactive.
15) Sketch some molecules that could be in a compound.
16) In what way is iron sulfide different from a mixture of iron and sulfur?
17) Is it easy to split a compound back up into its original elements?
18) Write down the two rules for naming compounds.
19) If two atoms of the same element combine, what happens to their name?
20)*Give the name of the following: a) MgO b) CaO c) NaCl d) $CaCO_3$ e) $CuSO_4$
21)*Give the name of the compound you get from chemically joining up these:
    a) sodium with chlorine b) magnesium with chlorine c) magnesium with carbon and oxygen.
22) What is a pure substance? What is a mixture?
23) Describe what happens when a substance dissolves.
24) List four mixture separation techniques with an example for each one.
25) Which of them would you use to try to identify different colours in a paint?
26) List the 12 facts you need to know about metals.
    Then list the 9 facts you need to know about non-metals.
27) What are ceramics useful for?
28) Name a composite material and describe what it's made of.

*Answers on page 193

Chemical Reactions

In a chemical reaction, all that's really happening is the atoms moving around into new formations. The reactants might get hotter or give off coloured light or thick smoke, but their total mass won't change.

Atoms Rearrange Themselves in a Chemical Reaction

1) In a chemical reaction atoms are not created or destroyed.
2) The atoms at the start of a reaction are still there at the end.
3) Bonds get broken and made in the reaction, as atoms rearrange themselves in going from the reactants to the products. But the atoms themselves are not altered.

EXAMPLE: zinc + copper sulfate ⟶ zinc sulfate + copper

Zn + Cu S ⟶ Zn S + Cu

The Mass Doesn't Change in a Chemical Reaction

1) In a chemical reaction no mass is lost or gained when the reactants turn into the products.
2) This is because the total number of atoms is the same before and after the reaction.
3) Chemical reactions involve a change in energy, i.e. reactions always give out or take in energy (p.77). This energy is usually transferred by heating, which causes the temperature in a reaction to go up or down.
4) Visible changes can occur in the reaction mixture. These show that a reaction has taken place. For example — a gas comes off, a solid is made, or the colour changes.

EXAMPLE: When magnesium reacts with blue copper sulfate solution, the solution goes colourless, copper coats the magnesium strip and the temperature rises. But the mass stays the same.

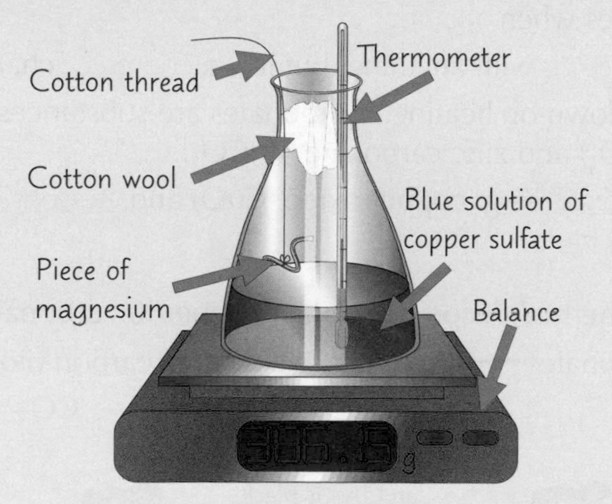

Chemical reactions — just a case of atoms moving around

Some chemical reactions involve colour changes, heating up, stinky emissions and even explosions, but there's one thing that always stays the same — the total mass, before and after the reaction.

Examples of Chemical Reactions

This page contains three common examples of chemical reactions.

Combustion is Burning in Oxygen

1) Combustion is burning — a fuel reacts with oxygen to release energy.
2) Three things are needed for combustion:

   1) Fuel
   2) Heating
   3) Oxygen

3) Hydrocarbons are fuels containing only hydrogen and carbon. When it's hot enough and there's enough oxygen, hydrocarbons combust (burn) to give water and carbon dioxide:

hydrocarbon + oxygen ⟶ carbon dioxide + water (+ energy)

4) Combustion is useful because energy is transferred away by heating and light. It's the process behind candles, wood fires, car engines, coal power plants, etc.

Oxidation is the Gain of Oxygen

1) When a substance reacts and combines with oxygen, it's called an oxidation reaction.
2) Combustion is an oxidation reaction.
3) Another example of oxidation is rusting. Iron reacts with oxygen in the air to form iron oxide, i.e. rust.

iron + oxygen ⟶ iron oxide (rust)

Thermal Decomposition is Breaking Down when Heated

1) Thermal decomposition is when a substance breaks down into at least two other substances when heated.
2) The substance isn't actually reacting with anything, but it is a chemical change.
3) Some metal carbonates break down on heating. Carbonates are substances with CO_3 in them, like copper(II) carbonate ($CuCO_3$) and zinc carbonate ($ZnCO_3$).
4) They break down into a metal oxide (e.g. copper oxide, CuO) and carbon dioxide. This usually results in a colour change.

EXAMPLE: The thermal decomposition of copper(II) carbonate.

copper(II) carbonate ⟶ copper(II) oxide + carbon dioxide

$CuCO_3 \longrightarrow CuO + CO_2$

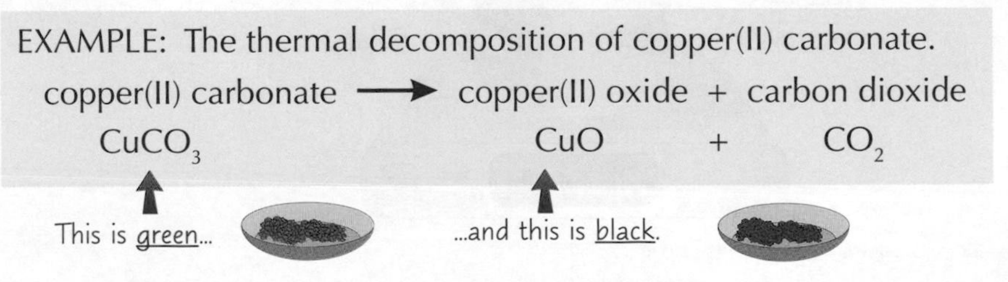

Thermal decomposition doesn't need any other reactants

Here are three common types of chemical reaction to read about and learn. A good way to learn key phrases and equations is to use flash cards — create some for this topic.

Chemical Reactions and Catalysts

Chemical reactions always involve a transfer of energy and catalysts can speed this transfer up.

In an **Exothermic** Reaction, **Energy** is **Transferred Out**

An exothermic reaction is one which transfers energy to the surroundings.

1) Energy is usually given out by heating, so exothermic reactions involve a rise in temperature.
2) The best example of an exothermic reaction is combustion (see previous page). This gives out a lot of energy — it's very exothermic.
3) Many neutralisation reactions (page 81) and oxidation reactions (previous page) are exothermic.
4) Everyday uses of exothermic reactions include hand warmers and self-heating cans of coffee.

In an **Endothermic** Reaction, **Energy** is **Taken In**

An endothermic reaction is one where energy is taken in from the surroundings.

1) Energy is usually taken in by heating, so endothermic reactions involve a fall in temperature.
2) Endothermic reactions are much less common. Thermal decompositions (previous page) are a good example, since they involve a substance taking in energy and breaking down.
3) Everyday uses of endothermic reactions include sports injury packs. They take in energy and get very cold.

Catalysts Increase the **Speed** of a Reaction

1) A catalyst is a substance which speeds up a chemical reaction, without being chemically changed or used up in the reaction itself.
2) Catalysts come out of a reaction the same as when they went in — usually they just give the reacting particles somewhere to meet up and do the business. That means catalysts can be reused.
3) Chemical reactions need energy to get them started — usually through heating. Catalysts lower the minimum amount of energy needed for a reaction to happen.
4) This means a lower temperature can be used to carry out the reaction.

> 1) Catalysts are very important for business — most industrial reactions use them.
> 2) By increasing the speed of the reaction and lowering the temperature needed, they make industrial reactions cheaper and increase the amount of product made in a given time.
> 3) There are some disadvantages to catalysts. They can be expensive to buy, and different reactions use different catalysts, so businesses can't get away with just buying one to use for everything.
> 4) They also need to be cleaned and they can be 'poisoned' by impurities.

"exo-" means out and "endo-" means in

"Exothermic" and "endothermic" sound pretty similar, but don't get them confused — they have very different meanings. Most reactions you'll see in the chemistry lab are exothermic, and they're usually the ones that are easier to spot — a flame is a dead giveaway.

Balancing Equations

It's important to live a balanced life — that includes work, play, nutrition and chemical equations.

Chemical Equations Show What Happens in a Reaction

You can show what happens in a chemical reaction using:

1) A WORD EQUATION — where the names of the products and reactants are written out in full.
2) A SYMBOL EQUATION — which uses chemical symbols and formulae.
   A balanced symbol equation shows how many of each chemical react or are made in a reaction.

Chemical Equations are Equal on Both Sides

Here's an example of writing a balanced equation for burning magnesium in oxygen.

1) Write the word equation: magnesium + oxygen ⟶ magnesium oxide
2) Write in the chemical formulae of all the reactants and products: $Mg + O_2 \longrightarrow MgO$
3) Check that the equation is balanced by counting the number of each atom on both sides of the equation. Then do steps A, B, C and D below to balance the atoms up one by one. Keep track of the number of atoms on each side as you go:

Left side of equation	Right side of equation
One Magnesium	One Magnesium
Two Oxygen	One Oxygen

A Find an element that doesn't balance and pencil in a number to try and sort it out.

There isn't enough oxygen on the right side of the equation — add "2" before MgO.

$$Mg + O_2 \longrightarrow 2MgO$$

B See where that gets you by counting up the atoms again.

Left side of equation	Right side of equation
One Magnesium	Two Magnesium
Two Oxygen	Two Oxygen

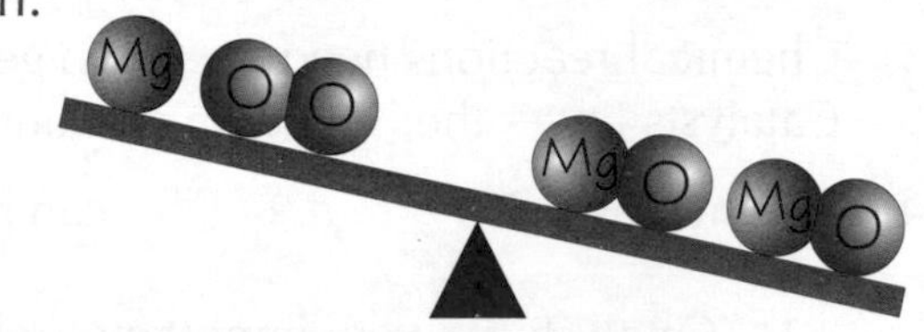

C Continue to chase the unbalanced atoms by going back to A) — pencil in a number before a formula, then see where it gets you when you count up the atoms.

There isn't enough magnesium on the left side of the equation — add a "2" before Mg.

$$2Mg + O_2 \longrightarrow 2MgO$$

D See where that gets you by counting up the atoms again.

Left side of equation	Right side of equation
Two Magnesium	Two Magnesium
Two Oxygen	Two Oxygen

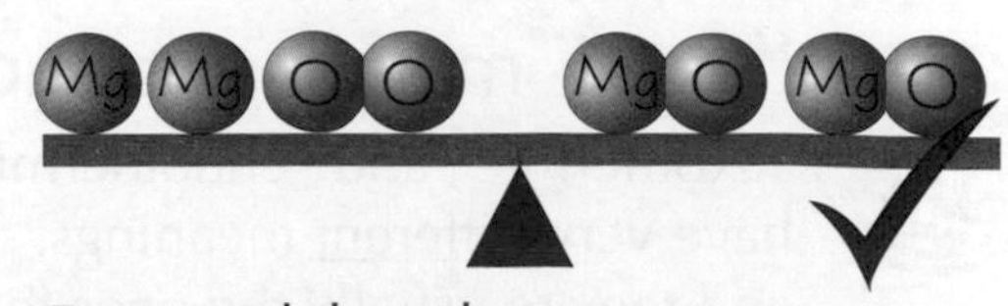

Done and dusted.

Warm-Up and Practice Questions

You need to learn all the stuff in this section. Might as well make a start on it now.
It's not a lot of fun, but it's the only way to get good marks.

Warm-Up Questions

1) 92 g of sodium reacts with 142 g of chlorine to make sodium chloride.
Calculate the total mass of the sodium chloride produced from this reaction.

2) List three visible changes in a reaction mixture which could show that a chemical change had taken place.

3) Is thermal decomposition an exothermic or an endothermic reaction? Explain why.

4) Explain why a catalyst can be used more than once.

5) Give two reasons why a chemical production plant might not want to use catalysts.

6) Write a balanced symbol equation for:

methane (CH_4) + oxygen $\longrightarrow$ carbon dioxide + water

Practice Questions

1 Chemical reactions can be endothermic or exothermic.

(a) Copy and complete the sentences below about **endothermic** reactions.
In an endothermic reaction, energy is .... the surroundings.
This is often shown by a .... in temperature .

(2 marks)

(b) Combustion and oxidation are both types of **exothermic** reaction.

(i) What is an exothermic reaction?

(1 mark)

(ii) State the three things needed for combustion to take place.

(1 mark)

(iii) Write a word equation for the oxidation reaction between iron and oxygen.

(1 mark)

2 Channing heats some copper carbonate ($CuCO_3$),
to produce copper oxide (CuO) and carbon dioxide.

(a) What type of reaction is this?

(1 mark)

(b) Suggest two ways in which Channing could tell that a reaction has taken place.

(2 marks)

(c) Write a balanced symbol equation for this reaction.

(2 marks)

(d) Channing wants to increase the amount of copper oxide produced for a given mass of copper carbonate. He decides to use a catalyst.
Explain why this won't have the effect he wants.

(1 mark)

Acids and Alkalis

The pH scale is what scientists use to describe how acidic or alkaline a substance is. Universal indicator takes on a colour based on the pH of the substance it's mixed with.

The pH Scale Shows the Strength of Acids and Alkalis

1) The pH scale goes from 0 to 14.
2) pH can also be shown by the colour universal indicator turns (see below).
3) The scale tells you how strongly acidic or alkaline a substance is (or whether they're neutral).
4) Anything with a pH below 7 is an acid and will turn universal indicator red, orange or yellow.
5) The strongest acids have pH 0.
6) Anything with a pH above 7 is an alkali and will turn universal indicator blue or purple.
7) The strongest alkalis have pH 14.
8) A neutral substance (e.g. pure water) has pH 7 and will turn universal indicator green.

pH 0 1 2 3 4 5 6 7 8 9 10 11 12 13 14

Strong ACIDS — Weak ACIDS | Weak ALKALIS — Strong ALKALIS

NEUTRAL

Indicators Are Special Dyes Which Change Colour

1) An indicator is just something that changes colour depending on whether it's in an acid or in an alkali.
2) Litmus paper is quite a popular indicator, but it only tells us whether a liquid is an acid or an alkali — it does not say how strong it is. Acids turn litmus paper red and alkalis turn it blue.
3) Universal indicator solution is a very useful mixture of dyes which gives the colours shown in a pH chart.

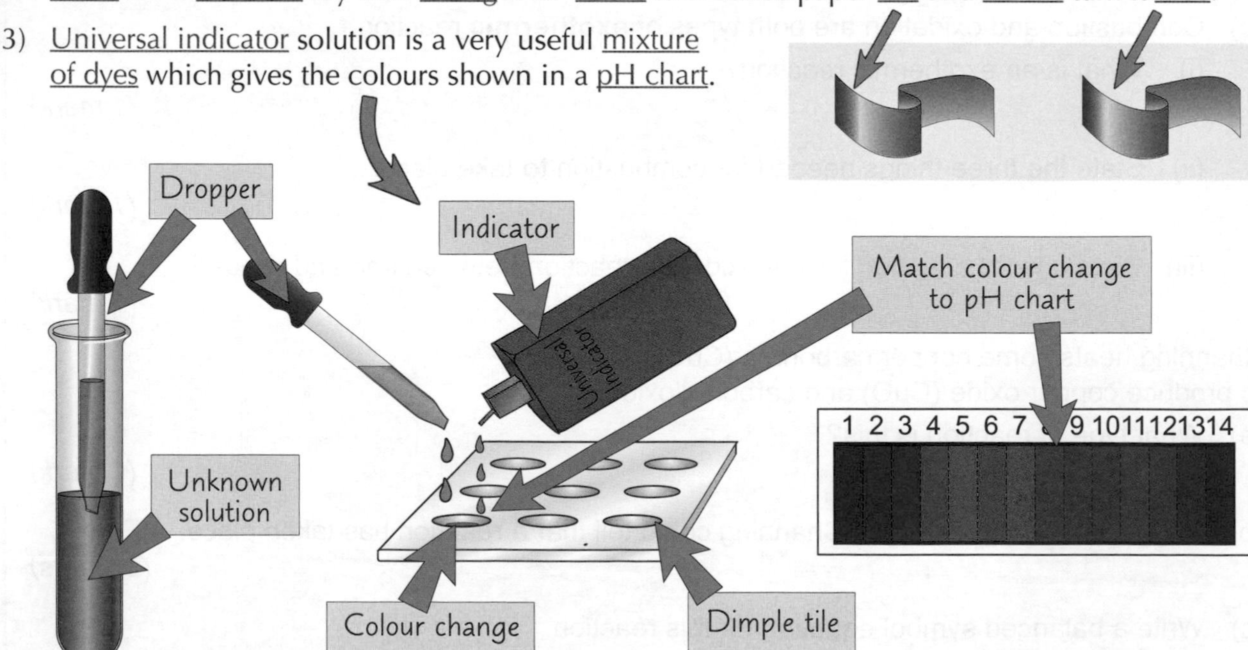

We use loads of acids and alkalis in our daily life

For example, vinegar has a pH of about 2, and even milk is a bit acidic (about 6.5). Lots of cleaning products are alkalis. Washing-up liquid is slightly alkaline (pH 7.5-8) and bleach has a pH of about 11.

Neutralisation Reactions

You might have done something like this in the lab and, if not, I bet you will pretty soon.
Make sure you know all this stuff — it's pretty easy and a super-useful thing to know about.

Acids and Alkalis Neutralise Each Other

1) Acids react with alkalis to form a neutral solution of a salt and water:

acid + alkali ⟶ salt + water

2) This is known as a neutralisation reaction because the products have a neutral pH, i.e. a pH of 7.

Making Salts by Neutralisation

Making salts is pretty easy — you just need a steady hand and a lot of time.

Hydrochloric acid

acid added in drops to neutralise

25 cm^3 of sodium hydroxide solution

glass rod

solution

Universal indicator paper

test if solution is neutral every few drops

boil off some of the water

leave to evaporate until sodium chloride salt crystals form

1) Wearing eye protection, add an acid to an alkali dropwise with a pipette.
2) After every few drops, remove a small sample to check if the pH is neutral (pH 7).
3) Keep adding acid until the solution is neutral.
4) When it's neutral the solution is put in an evaporating dish and about two thirds of it can be boiled off to make a saturated solution of the salt.
5) Leave this solution overnight for the rest of the water to evaporate and nice big salt crystals will form. The slower the crystallisation, the bigger the crystals.

A saturated salt solution can't have any more salt dissolved in it. See page 61.

To Change the Salt, You Must Change the Acid

1) The salt you get out of the neutralisation reaction above depends on the acid you use.
2) The clue is normally in the name:

Hydrochloric acid reacts to make chloride salts... like sodium chloride.

Sulfuric acid reacts to make sulfate salts... like copper sulfate.

Nitric acid reacts to make nitrate salts... like sodium nitrate.

Warm-Up and Practice Questions

These questions are designed to test what you know. If you've gone through the section and learnt the facts, then you should be able to have a go at these questions.

Warm-Up Questions

1) What number range does the pH scale cover?
2) What pH might an acid have? What pH might an alkali have?
3) What is an indicator?
4) Why is litmus paper not as useful as Universal Indicator?
5) What kind of acid would you use to make a sulfate salt?

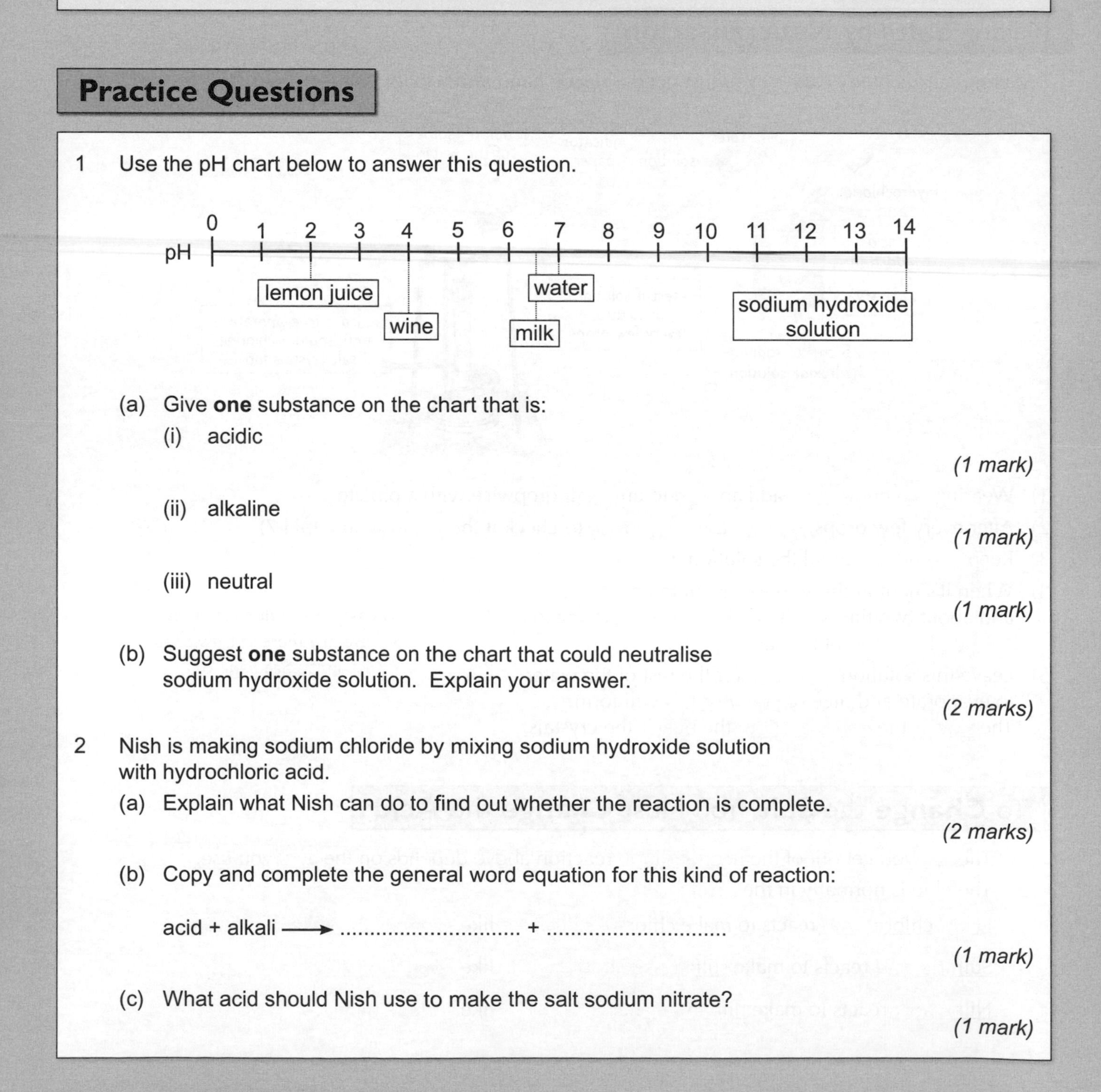

Practice Questions

1 Use the pH chart below to answer this question.

(a) Give **one** substance on the chart that is:

(i) acidic

(1 mark)

(ii) alkaline

(1 mark)

(iii) neutral

(1 mark)

(b) Suggest **one** substance on the chart that could neutralise sodium hydroxide solution. Explain your answer.

(2 marks)

2 Nish is making sodium chloride by mixing sodium hydroxide solution with hydrochloric acid.

(a) Explain what Nish can do to find out whether the reaction is complete.

(2 marks)

(b) Copy and complete the general word equation for this kind of reaction:

acid + alkali ⟶ .... + ...

(1 mark)

(c) What acid should Nish use to make the salt sodium nitrate?

(1 mark)

Reactivity Series and Metal Extraction

You need to know which metals are most reactive — and which are least reactive.

The **Reactivity Series** — How **Well** a Metal **Reacts**

The Reactivity Series lists metals in order of their reactivity towards other substances.

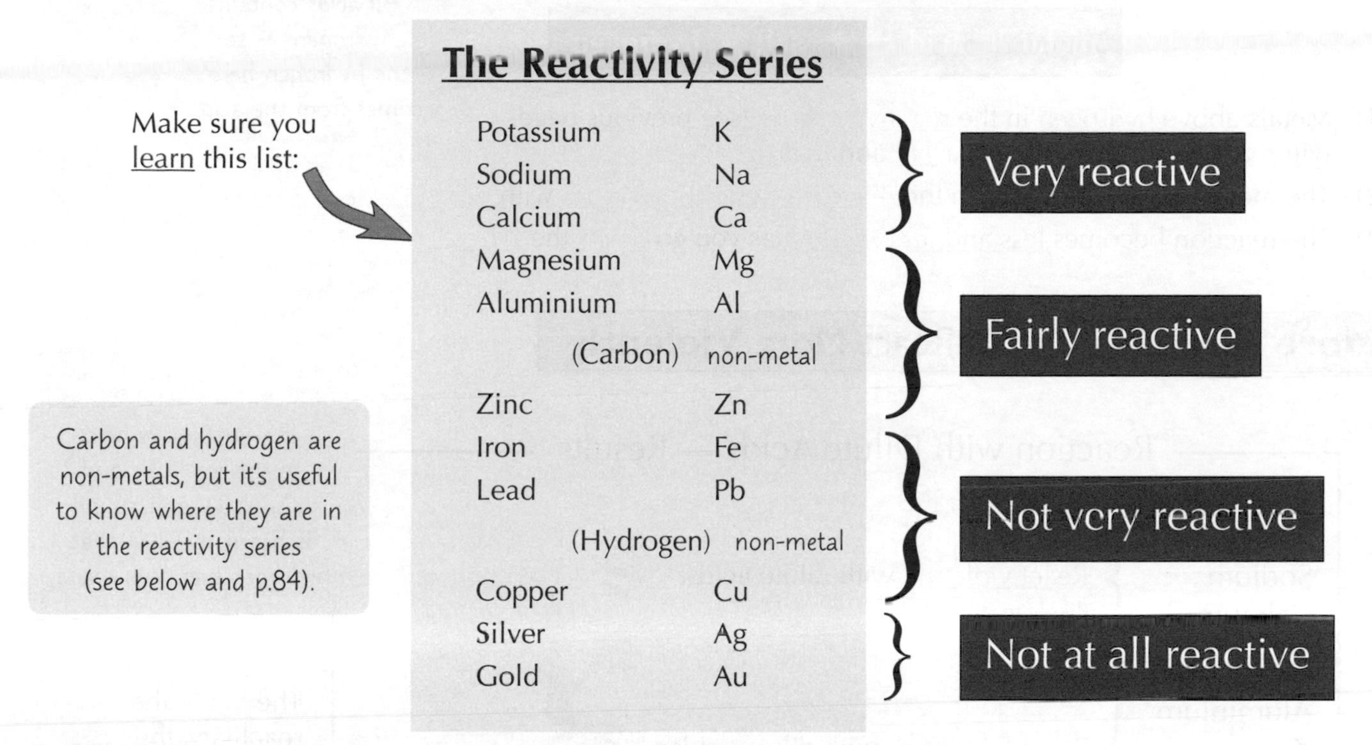

Some **Metals** Can Be **Extracted** with **Carbon**

1) Metals are usually mined as ores — rocks containing different metals and metal compounds (usually metal oxides — see page 85).
2) A metal can be extracted from its ore by reduction using carbon. When an ore is reduced, oxygen is removed from it. E.g. the oxygen is removed from iron oxide to extract the iron:

iron oxide + carbon ⟶ iron + carbon dioxide

3) Only metals that are less reactive than carbon (i.e. metals below carbon in the reactivity series) can be extracted from their ore using carbon.
4) Metals that are more reactive than carbon need to be extracted using electrolysis (where electricity splits up the ore into the elements that make it up).
5) Some metals, like silver and gold, are pretty unreactive, so they're often found in their pure form.

Potassium
Sodium
Calcium
Magnesium
Aluminium
—CARBON—
Zinc
Iron
Lead
Copper
Silver
Gold

A metal's reactivity shows how it will behave in reactions

In chemistry, the reactivity of a metal is an important feature. A phrase to help you learn the series is Peter Stopped Calling Me After Certain Zany Individuals Lost His Cool Sun Glasses.

Reaction of Metals with Acids

One more page on metals to learn. You don't need to know about each individual reaction, just how the reactivity of each metal affects it.

Reacting Metals with **Dilute Acid**

metal + acid ⟶ salt + hydrogen

All acids contain hydrogen — so the hydrogen here comes from the acid.

1) Metals above hydrogen in the reactivity series (see previous page) will react with acids to make a salt and hydrogen.
2) The metals below hydrogen in the reactivity series don't react with acids.
3) The reaction becomes less and less exciting as you go down the series.

More Reactive Metals React More **Violently**

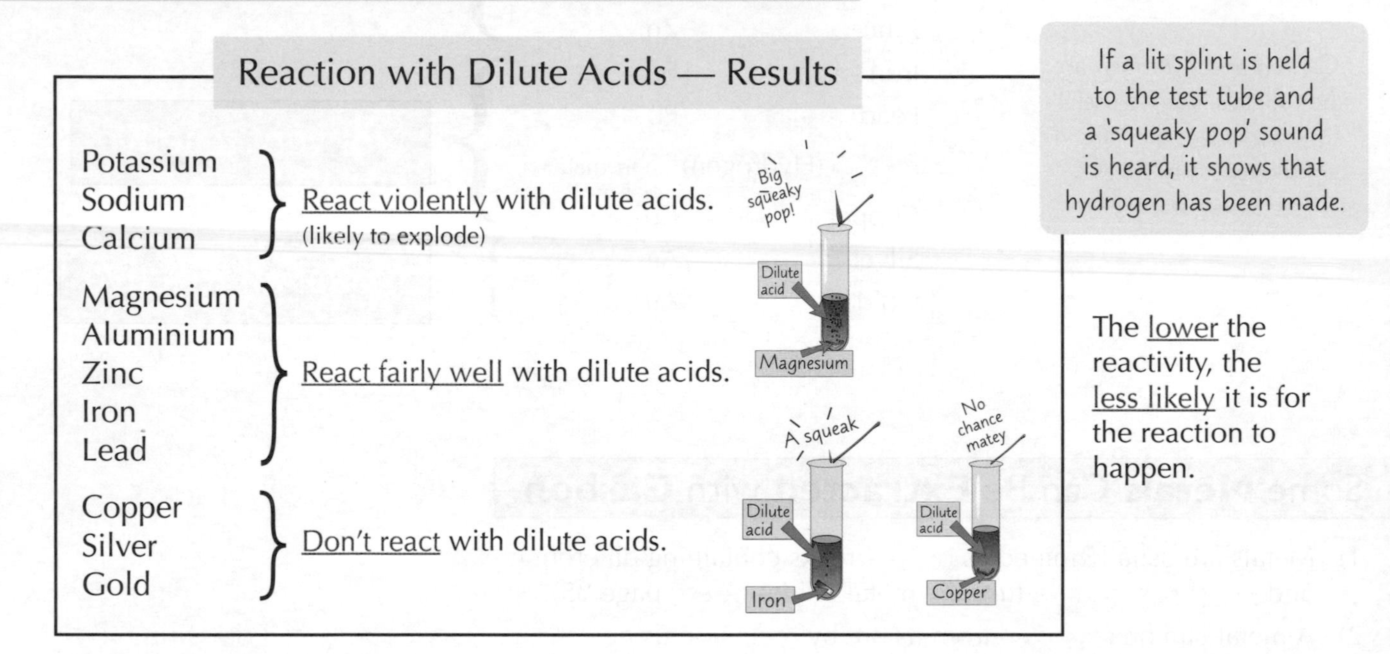

The lower the reactivity, the less likely it is for the reaction to happen.

EXAMPLES:

a) zinc + sulfuric acid ⟶ zinc sulfate + hydrogen

$$Zn + H_2SO_4 \longrightarrow ZnSO_4 + H_2$$

The zinc takes the place of the hydrogen in the acid because it's more reactive than the hydrogen.

b) sodium + hydrochloric acid ⟶ sodium chloride + hydrogen

$$2Na + 2HCl \longrightarrow 2NaCl + H_2$$

The sodium takes the place of the hydrogen in the acid — again because it's more reactive than the hydrogen.

The more reactive the metal, the more violent the reaction

It might seem like there's loads going on here, but it's just the same principle repeated over and over. All the metals that react have roughly the same reaction — some are just more violent than others.

Reactions of Oxides with Acids

Oxides are pretty self-explanatory — they've got oxygen in them somewhere.

Metals React with **Oxygen** to Make **Oxides**

Metals react with oxygen to make metal oxides.

E.g. magnesium + oxygen → magnesium oxide.

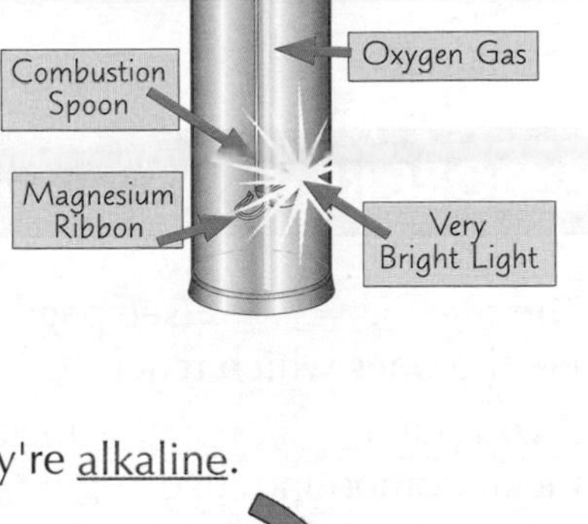

Metal Oxides are **Alkaline**

1) Metal oxides in solution have a pH which is higher than 7 — i.e. they're alkaline.
2) So metal oxides react with acids to make a salt and water.

acid + metal oxide ⟶ salt + water

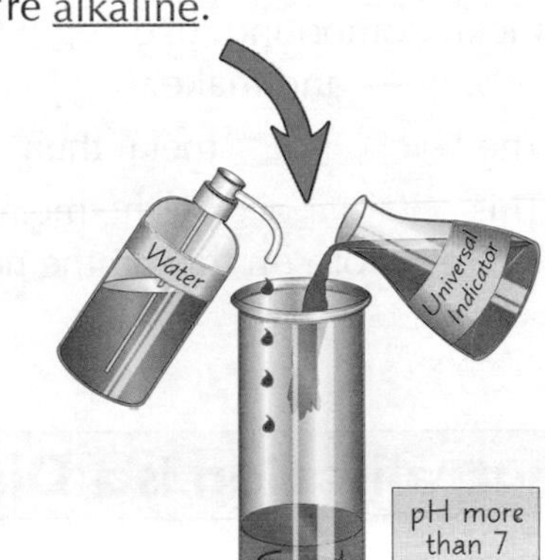

EXAMPLES:

hydrochloric acid + copper oxide → copper chloride + water

sulfuric acid + zinc oxide → zinc sulfate + water

nitric acid + magnesium oxide → magnesium nitrate + water

Non-metals React with **Oxygen** to Make **Oxides**

Non-metals also react with oxygen to make oxides.

E.g. sulfur + oxygen → sulfur dioxide.

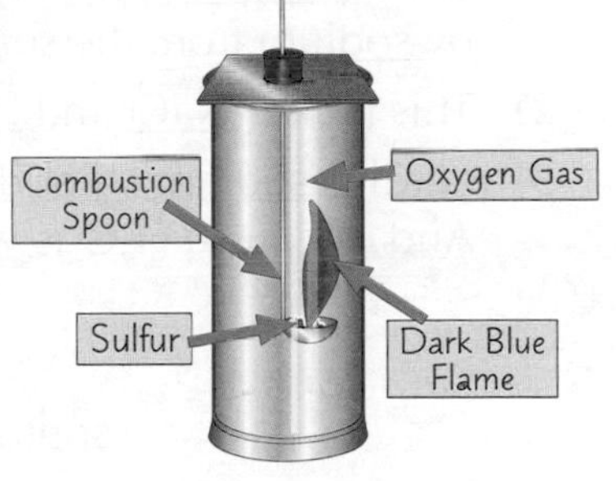

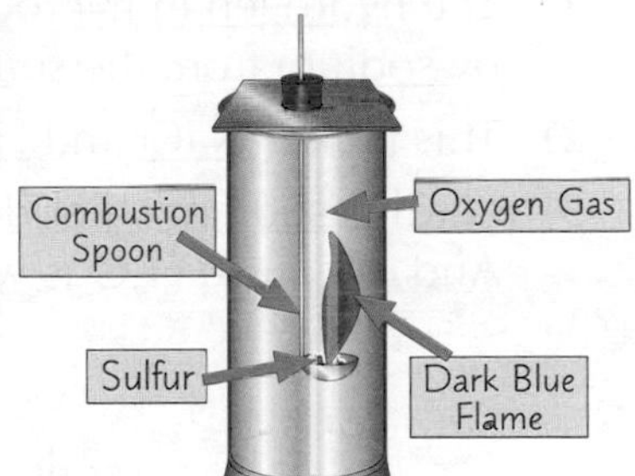

Non-metal Oxides are **Acidic**

1) The oxides of non-metals have a pH below 7. This means they're acidic.
2) So non-metal oxides will react with alkalis to make a salt and water.

alkali + non-metal oxide ⟶ salt + water

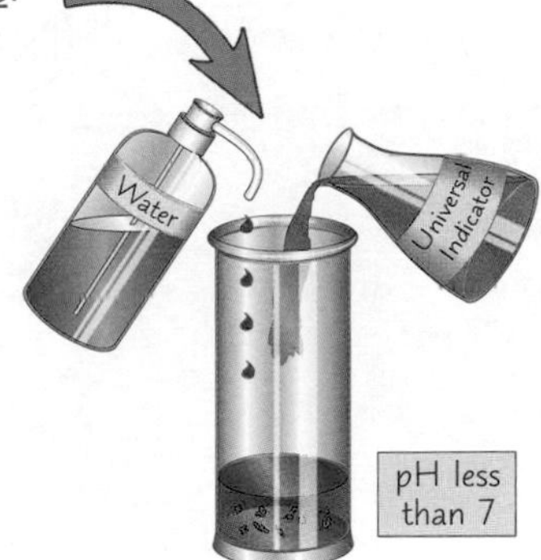

EXAMPLE:

sodium hydroxide + silicon dioxide → sodium silicate + water

an alkali (sodium hydroxide); a non-metal oxide (silicon dioxide)

REVISION TIP

Metal oxides are alkaline, non-metal oxides are acidic

Remember this key fact, and the word equation acid + alkali → salt + water (p.81). Then you can work out the word equations from this page for yourself (by sticking in either 'metal oxide' or 'non-metal oxide'), instead of having to remember three separate equations.

Displacement Reactions

You can use the reactivity series to work out if one metal will displace another...

'Displacement' Means 'Taking the Place of'

Learn this rule:

A MORE REACTIVE metal will displace a LESS REACTIVE metal from its compound.

1) The reactivity series (see page 83) tells you which are the most reactive metals — i.e. the ones which react most strongly with other things.
2) If you put a more reactive metal like magnesium into a solution of a less reactive metal compound, like copper sulfate, then magnesium will take the place of the copper — and make magnesium sulfate.
3) The "kicked out" metal then coats itself on the reactive metal, so we'd see copper.
4) This only happens if the metal added is more reactive — higher displaces lower. Got it? There's more on this on the next page.

Neutralisation is a Displacement Reaction

The reaction between hydrochloric acid and sodium hydroxide (to make sodium chloride) is also a displacement reaction...

1) The hydrogen in hydrochloric acid is displaced (or replaced) by sodium from the sodium hydroxide (the alkali).
2) This makes NaCl and H_2O.
3) NaCl is sodium chloride — common salt. And of course H_2O is water.

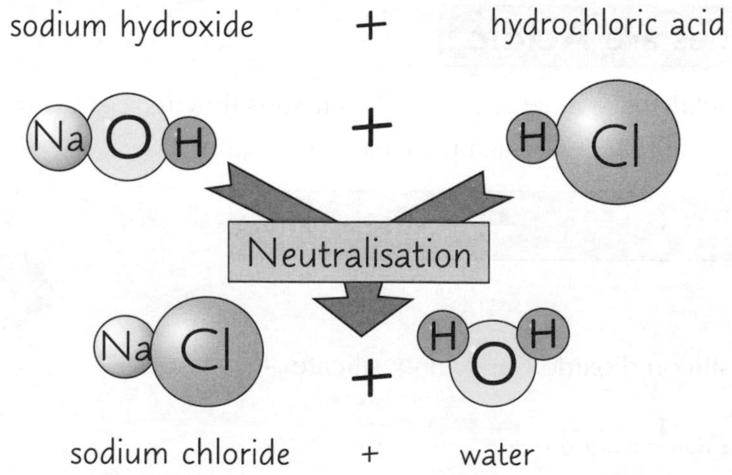

sodium hydroxide + hydrochloric acid

Na O H + H Cl

Neutralisation

Na Cl + H O H

sodium chloride + water

More reactive metals displace less reactive metals

Displacement reactions are really important, so make sure you learn them properly. Metals higher in the reactivity series displace those lower down. You'll see this effect in action on the next page.

Displacement Reactions

You can investigate the reactivity series of metals by experimenting with displacement reactions.

A **Reactivity Series** Investigation

Method: Put a bit of metal into some salt solutions and see what happens.

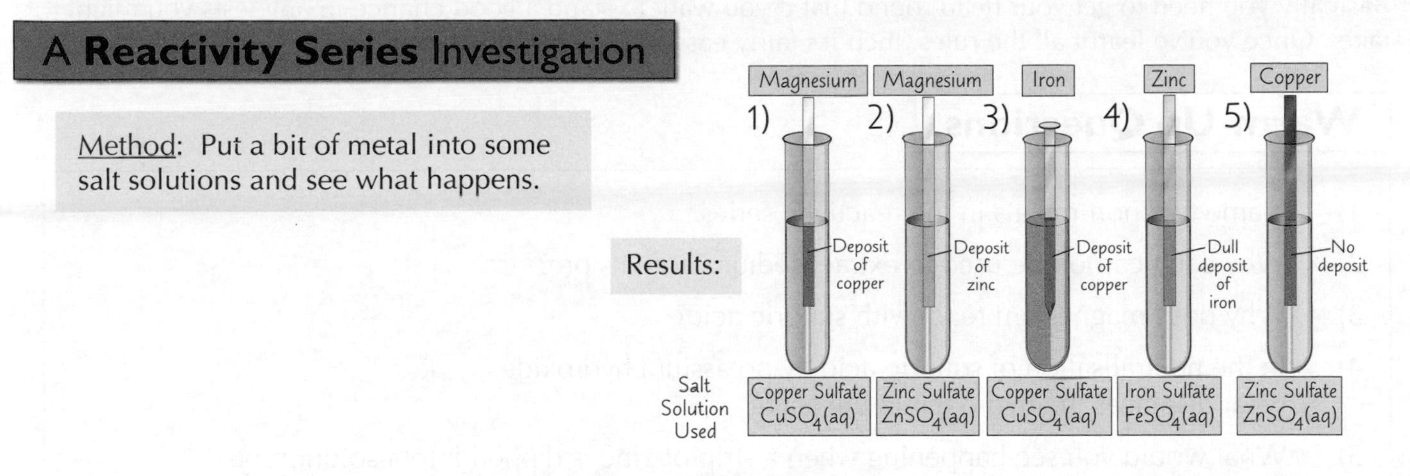

Tube 1: The blue copper sulfate solution goes colourless and the copper coats the magnesium strip. Magnesium must be more reactive than copper as it takes its place.

magnesium + copper sulfate ⟶ magnesium sulfate + copper

Tube 2: Zinc is seen coating the magnesium strip. Magnesium must be more reactive than zinc as it takes its place.

magnesium + zinc sulfate ⟶ magnesium sulfate + zinc

Tube 3: The blue copper sulfate solution goes green and the copper coats the nail. Iron must be more reactive than copper as it takes its place.

iron + copper sulfate ⟶ iron sulfate + copper

Tube 4: Iron is seen coating the zinc strip. Zinc must be more reactive than iron as it takes its place.

zinc + iron sulfate ⟶ zinc sulfate + iron

Tube 5: There's no reaction. Copper can't displace zinc — it's not reactive enough.

copper + zinc sulfate ⟶ no way

Conclusion:

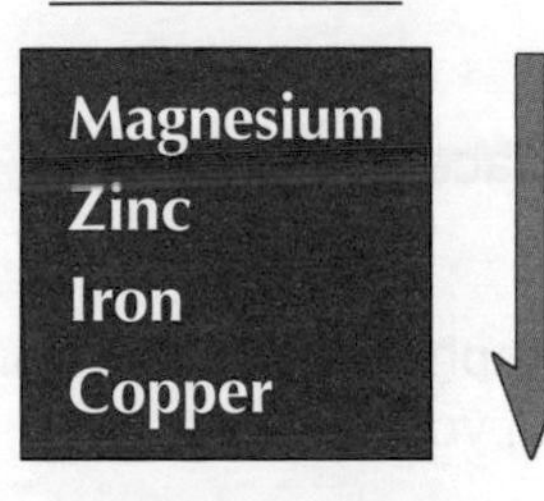

Least Reactive

Warm-Up and Practice Questions

The good thing about those last few pages is that they all revolve around one thing — the reactivity series. Basically you need to get your head round that if you want to stand a good chance in any tests you might take. Once you've learnt all the rules, then it's fairly easy to work out the answer to any question about it.

Warm-Up Questions

1) Name two non-metals in the reactivity series.
2) Why can't carbon be used to extract sodium from its ore?
3) Why does magnesium react with sulfuric acid?
4) In the neutralisation of sulfuric acid by potassium hydroxide, what displaces the hydrogen in the acid?
5) What would you see happening when a strip of zinc is dipped into a solution of:
   a) colourless magnesium sulfate solution?
   b) blue copper sulfate solution?

   Explain your answers.

Practice Questions

1 Stephen poured equal quantities of black copper oxide and grey zinc powders onto an upturned dish. He used a Bunsen burner to heat the mound of powder. A reaction started and Stephen removed the Bunsen burner. The mixture left behind was glowing, and Stephen saw a yellow solid which turned white when it was cold.

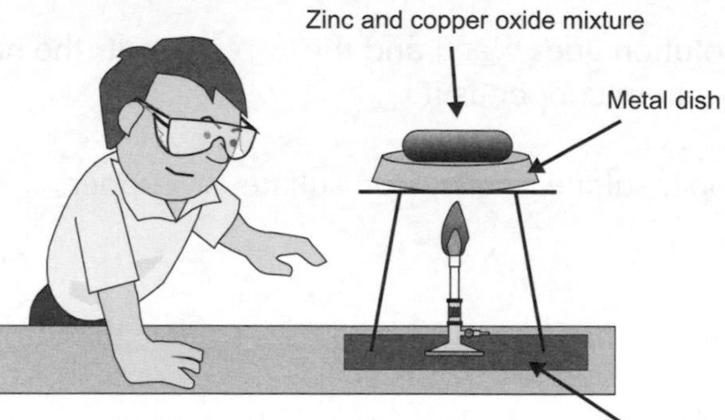

(a) Give **one** safety precaution which Stephen took during this experiment.

(1 mark)

(b) In this reaction, copper and zinc oxide were produced.

(i) Write a word equation to show this reaction.

(1 mark)

(ii) What kind of reaction is this?

(1 mark)

(iii) Why did this reaction take place?

(1 mark)

(c) What would have happened if Stephen had used aluminium oxide instead of copper oxide? Explain your answer.

(2 marks)

Practice Questions

2 The table below shows some reactions of different metals.

metal	reaction with oxygen	reaction with dilute acids
mercury	reacts slowly to form a red-orange solid.	no reaction
lithium	burns easily, with a bright red flame, to form a white solid	reacts quickly, releasing hydrogen gas
silver	no reaction	no reaction
magnesium	burns easily, with a bright white flame, to form a white solid	reacts slowly, releasing hydrogen gas

(a) Write the names of these four metals in the order of their reactivity, beginning with the most reactive.

(4 marks)

(b) Write a word equation for the reaction of magnesium with hydrochloric acid.

(1 mark)

The following table shows some reactions of different oxides.

oxide	reaction with acids	reaction with alkalis
potassium oxide	reacts, making a salt and water	no reaction
silicon dioxide	no reaction	reacts, making a salt and water

(c) A piece of red litmus paper is dipped into a solution containing one of the two oxides listed in the table. The paper turns blue. Which oxide is in the solution? Explain your answer.

(2 marks)

3 In the past, the only metals in use were ones that could be extracted from their ores using carbon reduction. This helps archaeologists to find out how old objects are. For example, if they find an object made of aluminium, they know the metal must be no more than a few centuries old.

(a) Explain why the metal in aluminium objects could not have been extracted using traditional carbon reduction methods.

(2 marks)

(b) Suggest one other metal that is not likely to have been in use a few centuries ago.

(1 mark)

(c) Explain why it is unlikely that any objects made from pure calcium would ever be found in the ground.

(1 mark)

Revision Summary for Section Six

There's no use getting through a whole section of Chemistry if you can't summarise it with a handy set of questions that test everything you need to know. Luckily for you, that's exactly what this page is for. You must have heard it all before by now, and it's the usual shtick — work through the questions one by one, make sure you know everything, then maybe treat yourself to something sweet.

1) What happens to the atoms in a chemical reaction? ☐
2) Does the mass change during a chemical reaction? Why or why not? ☐
3) What's combustion? ☐
4) What's the name of the process in which a chemical gains oxygen? ☐
5) What's thermal decomposition? ☐
6) What's formed when a metal carbonate breaks down by thermal decomposition? ☐
7) What's the main difference between exothermic and endothermic reactions? ☐
8)* When nitrogen and hydrogen combine to form ammonia, the temperature drops. Is this reaction exothermic or endothermic? ☐
9) How does a catalyst affect the speed of a reaction? ☐
10) Give two reasons why a chemical production plant might want to use catalysts. ☐
11)* Write a balanced symbol equation for: sulfur + oxygen ⟶ sulfur dioxide. ☐
12)* Write a balanced symbol equation for: calcium + oxygen ⟶ calcium oxide ☐
13) What pH does the strongest acid on a pH chart have? And the strongest alkali? ☐
14) What pH does a neutral solution have? ☐
15) What colour would universal indicator go if it was mixed with:
    a) a strong acid, b) a neutral solution, c) a strong alkali? ☐
16) What is neutralisation? ☐
17) Outline the method to make common salt — sodium chloride. ☐
18) Hydrochloric acid makes chloride salts — what salts does sulfuric acid make? ☐
19) What kind of salts do you get from nitric acid? ☐
20) List the reactivity series in the correct order. ☐
21) If you haven't already, add carbon and hydrogen to your reactivity series from the previous question. If you've already done this, give yourself a pat on the back. ☐
22) Which metals in the reactivity series can be extracted from their ores using carbon? Which can't? Explain why they can't. ☐
23) What do metals produce when they react with an acid? ☐
24) Which metal will react the most violently with acid? ☐
25) Are metal oxides in solution acidic, neutral or alkaline? ☐
26) Give an example of a neutralisation reaction involving a metal oxide. ☐
27) Are non-metal oxides in solution acidic, neutral or alkaline? ☐
28) What is the rule for displacement reactions? ☐
29) Explain why magnesium can displace copper from copper sulfate. ☐

*Answers on page 194.

The Earth's Structure

Ever wondered what the planet's like on the inside? This page will tell you all.

The Earth Has a **Crust**, a **Mantle** and a **Core**

The Earth is almost a sphere and it has a layered structure. A bit like a scotch egg. Or a peach.

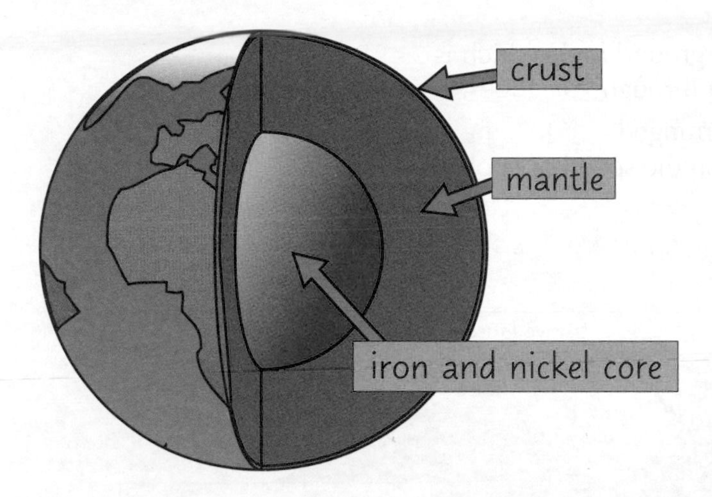

1) We live on the crust — a thin, outer layer of solid rock.
2) Below that is the mantle.
3) The mantle is mostly solid, but deep down it can flow very slowly (like a liquid). This is because the temperature increases as you go deeper into the mantle.
4) At the centre of the Earth is the core. We think it's made of iron and nickel.

The **Earth's Surface** is Made Up of **Tectonic Plates**

1) The crust and the upper part of the mantle are cracked into a number of large pieces. These pieces are called tectonic plates.
2) Tectonic plates are a bit like big rafts that 'float' on the mantle. They're able to move around.
3) The map below shows the edges of the plates as they are now, and the directions they're moving in (red arrows).
4) Most of the plates are moving very slowly (a few centimetres a year).
5) Sometimes, the plates move very suddenly, causing an earthquake.
6) Volcanoes and earthquakes often happen where two tectonic plates meet.

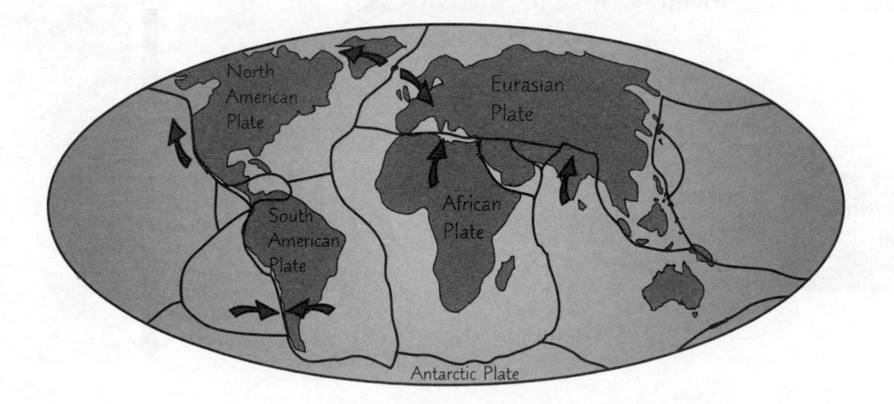

The part of Earth we live on is called the crust

You need to know the structure of Earth, i.e. what it would look like if you cut it open (which I wouldn't recommend) and what it's made of. That top diagram is your friend — learn it and learn it well. And, while we're on the subject, you'll need to learn all the words too. On the whole page.

Rock Types

There's more than one sort of rock you know — they're all covered on these two pages.

There are **Three Different** Types of **Rock**

1) **Igneous** Rocks

1) These are formed from magma (melted underground rock) which is pushed up through the crust — and often out through volcanoes.
2) They contain various minerals in randomly arranged interlocking crystals. The size of the crystals (or texture) depends on the speed of cooling. Large crystals mean that the rock has cooled slowly.
3) There are two types of igneous rocks: extrusive and intrusive.

EXAMPLES: basalt (extrusive), granite (intrusive).

Extrusive igneous rocks — cool quickly above ground.

Intrusive igneous rocks — cool slowly under ground and eventually get exposed when rocks above them wear away.

2) **Sedimentary** Rocks

1) These are formed from layers of sediment (rock fragments or dead matter) laid down in lakes or seas over millions of years. Sedimentary rocks can also form when water evaporates and leaves a dissolved solid (like salt) behind.
2) The layers are cemented together by other minerals.

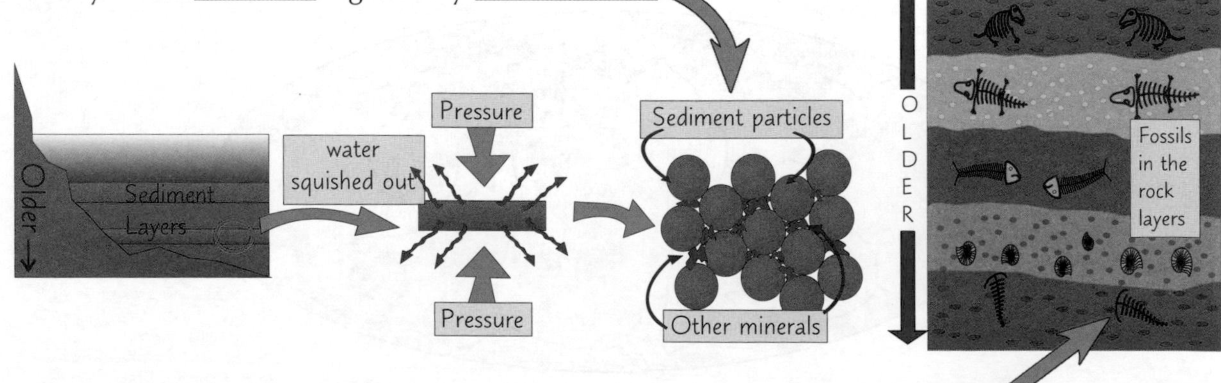

3) Fossils can form in the sediments. These are the long dead remains of plants and animals. The type of fossil is used to work out the relative age of the rock.

EXAMPLES: limestone, chalk, sandstone.

Rock Types

3) Metamorphic Rocks

1) These are the result of <u>heat</u> and <u>increased pressure</u> acting on existing rocks over <u>long</u> periods of time.
2) They may have really <u>tiny crystals</u> and some have layers.

<u>EXAMPLES</u>: marble, slate, schist.

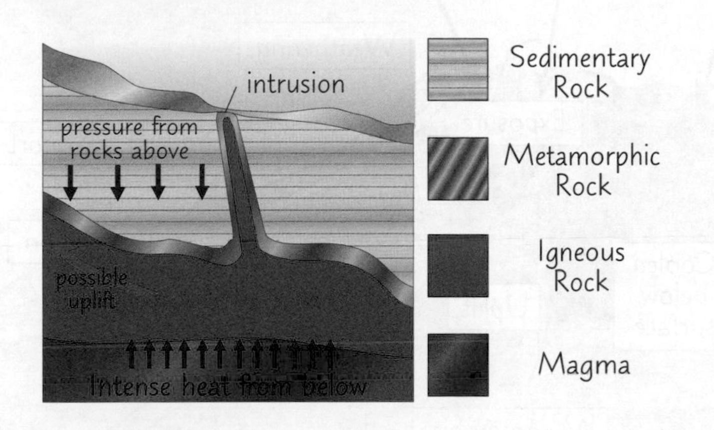

Rocks Are Made of **Minerals**

<u>Elements</u> and <u>compounds</u> make up <u>minerals</u> — and these make up <u>rocks</u> in the crust. E.g.

Elements	Compound	Mineral	Rock
Silicon & Oxygen	Silicon dioxide	Quartz	Granite

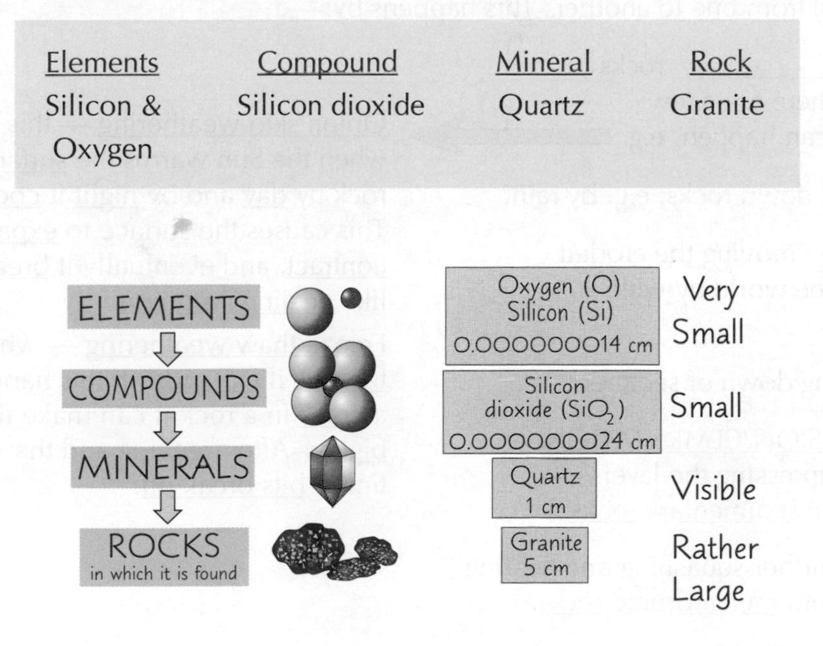

Some people think all rocks are the same, but they're wrong

<u>Memorise</u> the headings, then cover the pages and scribble down what you know.
Learn some <u>examples</u> of each type of rock while you're at it, and you'll soon be a geologist.

The Rock Cycle

The rock cycle involves changes to rocks both inside and outside the Earth.

The Rock Cycle Takes Millions of Years to Complete

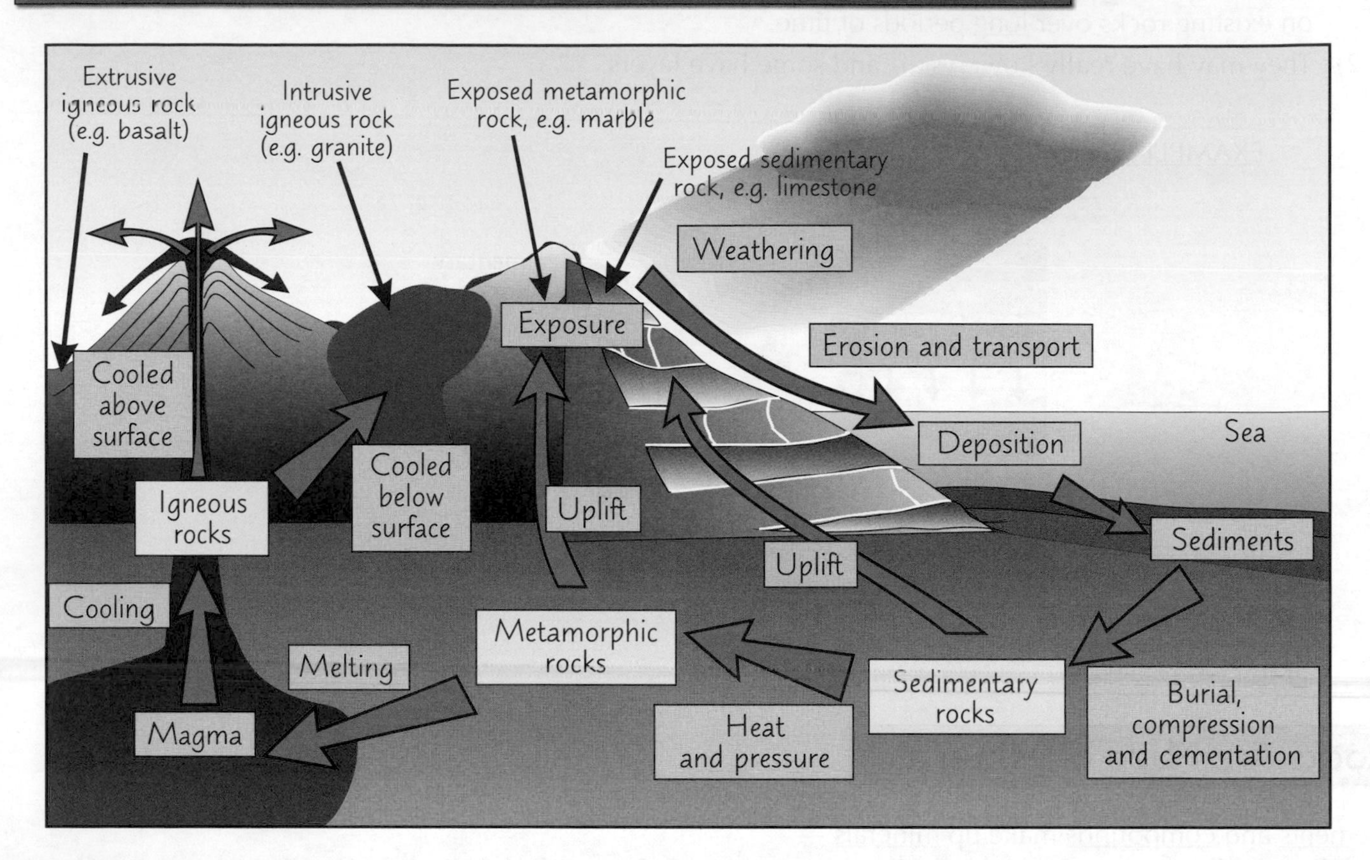

The rock cycle involves changing the three types of rock (igneous, sedimentary and metamorphic, see previous two pages) from one to another. This happens by:

1) WEATHERING: breaking down rocks into smaller bits. There are a few different ways this can happen, e.g. →

> Onion skin weathering — this happens when the Sun warms the surface of a rock by day and by night it cools down. This causes the surface to expand and contract, and eventually it breaks away, like peeling an onion.
>
> Freeze-thaw weathering — when water freezes, it expands. If this happens in a crack in a rock it can make the crack bigger. After freezing and thawing many times, bits break off.

2) EROSION: wearing down rocks, e.g. by rain.
3) TRANSPORTATION: moving the eroded bits of rock round the world by wind and water (mostly).
4) DEPOSITION: laying down of sediment.
5) BURIAL/COMPRESSION/CEMENTATION: squeezing and compressing the layers — eventually they form sedimentary rocks.
6) HEAT/PRESSURE: further squashing and heating — turns the rocks into metamorphic rocks.
7) MELTING: intense heating makes the rock partially melt — that changes it to magma.
8) COOLING: solidification of the molten (melted) rock to form igneous rocks.
9) EXPOSURE: back to weathering and erosion again. Simple huh.
(The amount of rock on the surface is always about the same, even though it's weathered away.)

Warm-Up and Practice Questions

You'll need to learn all the stuff in this section if you want to make it through Key Stage 3 Science. There were loads of technical words over the previous few pages, so you might need to put in that extra bit of time to get them in your head. The best way to check you've learnt them all is to try answering a few questions. Might as well make a start now. It's not a lot of fun, but it's the only way to succeed.

Warm-Up Questions

1) The Earth's crust is made of large segments that can move around. What are these called?
2) In which type of rock might you find fossils?
3) A sedimentary rock is buried and subjected to high heat and pressure. What type of rock is formed?
4) List the following in order of size, starting with the smallest: mineral, compound, element, rock.
5) Name the nine processes in the rock cycle.
6) Describe two types of weathering in rocks.

Practice Questions

1 The Earth is made up of three main layers.

a) Name the parts of the Earth labelled A-C in the diagram below.

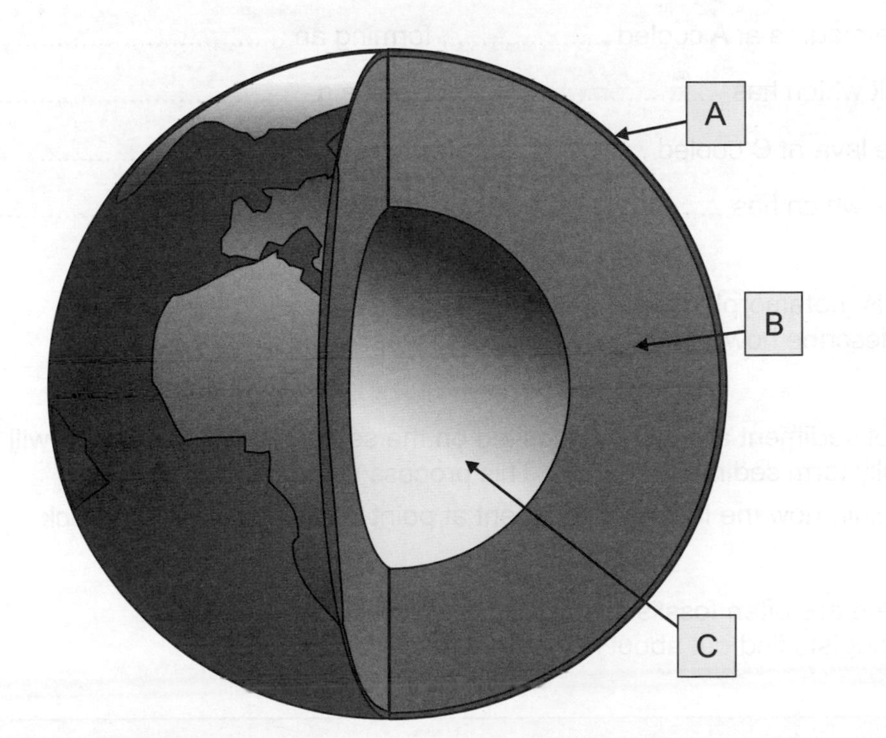

(3 marks)

b) Which layer is thought to be made of iron and nickel?

(1 mark)

Practice Questions

2 The diagram below shows part of a small volcanic island.
There have been no eruptions for thousands of years.

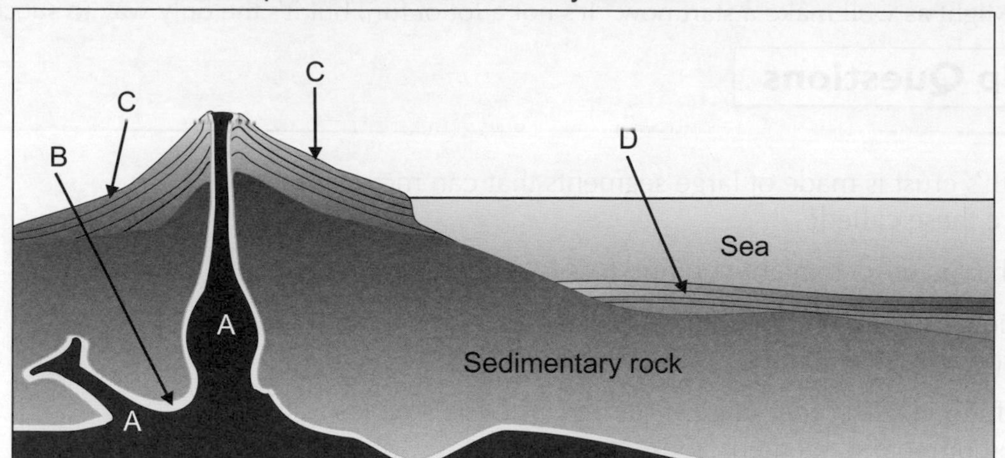

(a) Rocks A and C have both formed from solidifying magma.

(i) What type of rock is formed from solidifying magma?

(1 mark)

(ii) Copy and complete the sentences below using words from the box.

slowly	quickly	large	small	intrusive
extrusive	limestone	basalt	slate	granite

The magma at A cooled ...., forming an ....
rock which has .... crystals, e.g. ....
The lava at C cooled ...., forming an ....
rock which has .... crystals, e.g. ....

(8 marks)

(b) Rock B is metamorphic.
Briefly describe how this metamorphic rock was formed.

(2 marks)

(c) Layers of sediment are being deposited on the seabed at D. The layers will eventually form sedimentary rock. This process takes millions of years.

(i) Explain how the layers of sediment at point D will be turned into rock.

(2 marks)

(ii) There are often fossils in sedimentary rock. What can geologists find out about rocks from these fossils?

(1 mark)

Recycling

Every time you recycle something you're doing your bit to save limited resources.

The Earth is the Source of Almost All of Our Resources

1) For example, we get:

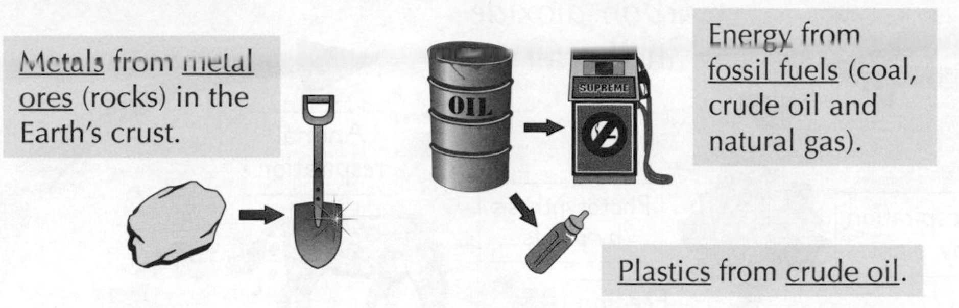

Fossil fuels are made from the remains of dead plants and animals buried in the Earth's crust for millions of years.

2) But these resources are limited. Once we've burnt all the Earth's fossil fuels or mined all the metal ores, that's it — we won't be getting any more any time soon. And that's where recycling comes in.

There are Lots of Good Reasons for Recycling

Recycling means taking old, unwanted products and using the materials to make new stuff.
Recycling is generally better than making things from scratch all the time because:

1) It uses less of the Earth's limited resources — things like crude oil and metal ores.
2) It uses less energy — which usually comes from burning fossil fuels.
3) Energy is expensive — so recycling tends to save money too.
4) It makes less rubbish — which would usually end up in landfill sites (rubbish dumps).

Example — recycling aluminium cans:

1) If aluminium wasn't recycled, more aluminium ore would have to be mined.
2) Mining costs money and uses loads of energy. It also makes a mess of the landscape.
3) The ore then needs to be transported and the aluminium extracted — which uses more energy.
4) It then costs to send the used aluminium to landfill.

It's a complex calculation, but for every 1 kg of aluminium cans that are recycled, you save:

- 95% of the energy needed to mine and extract 'fresh' aluminium,
- 4 kg of aluminium ore,
- a lot of waste.

It's really efficient to recycle aluminium.

It's usually more efficient (in terms of energy and cost) to recycle materials rather than throw them away and produce new ones. But the efficiency varies depending on what it is you're recycling. E.g. you get an energy saving of 95% by recycling aluminium, but less with plastics (70%) and steel (60%).

Recycle this book — but wait till you've finished KS3 science

Not all materials are as efficient to recycle as aluminium. But even if the energy and cost savings are relatively small, you could still be saving precious limited resources and creating less waste.

The Carbon Cycle

Carbon is a very important element because it's part of all living things.
As shown below, it's constantly recycled through the environment.

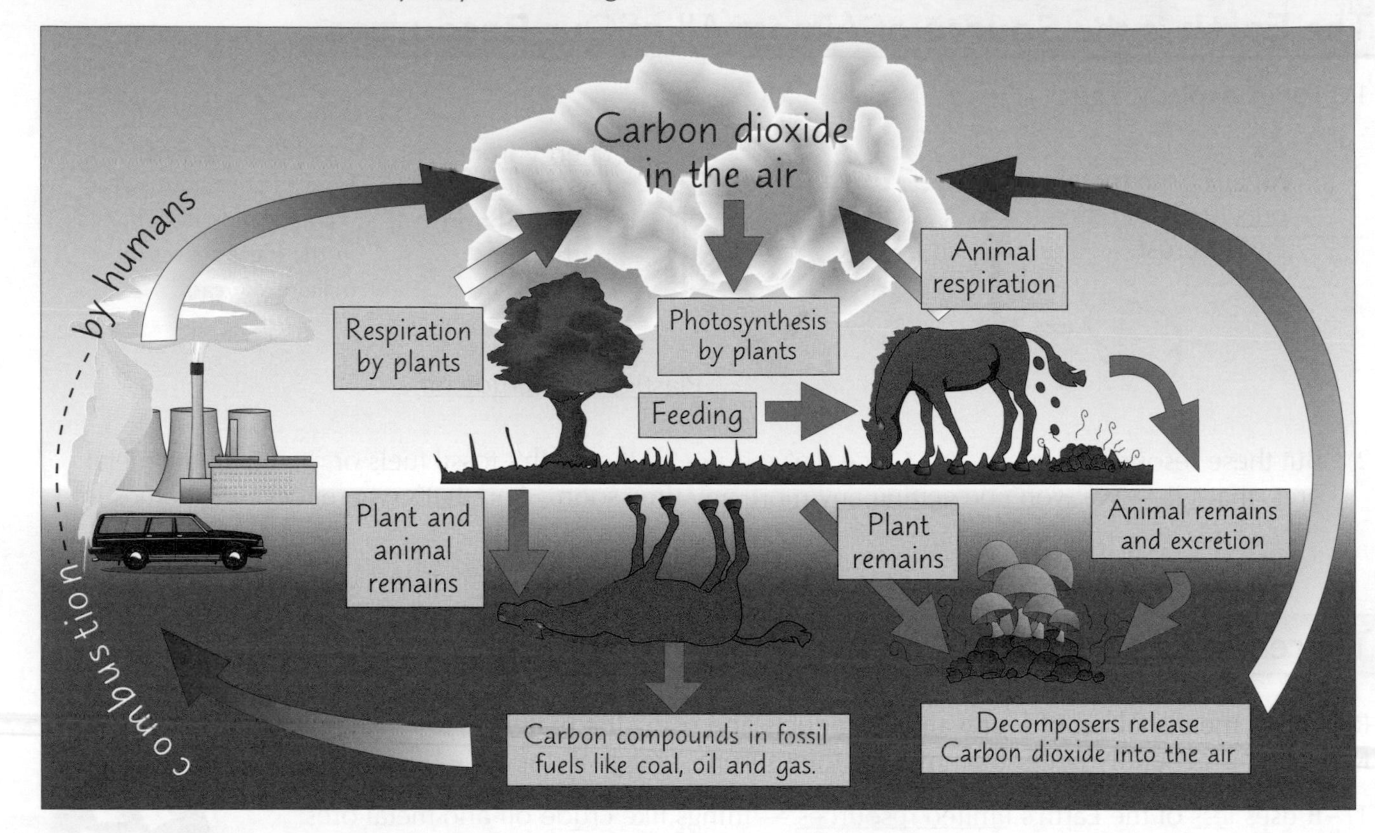

Learn these points:

1) **Photosynthesis** Removes **Carbon Dioxide** from the **Air**

1) Green plants and algae take in carbon dioxide from the air during photosynthesis (see p.30).
2) The plants and algae use the carbon to make carbohydrates, fats and proteins.

2) **Carbon** is **Passed Along** the **Food Chain** When Animals **Feed**

1) Some of the carbon in plants is passed on to animals when they eat the plants.
2) The animals then use the carbon to make fats and proteins of their own. The carbon moves along the food chain when the animals are eaten by other animals.

3) **Respiration** and **Combustion** Return **Carbon Dioxide** to the **Air**

1) Some carbon is returned to the air as carbon dioxide when plants and animals respire (see page 4).
2) When plants and animals die, decomposers (like bacteria and fungi) feed on them. Decomposers also feed on animal waste. When the decomposers respire, carbon dioxide is returned to the air.
3) Some dead plant and animal remains get buried and eventually form fossil fuels. When fossil fuels are burnt (combustion) this releases carbon dioxide back into the air.

The Atmosphere and the Climate

It's important to know exactly what you're breathing in and out. So read this page and find out.

The Earth's Atmosphere is Made Up of Different Gases

1) The gases that surround a planet make up that planet's atmosphere.
2) The Earth's atmosphere is around:

78% nitrogen (N_2) | 21% oxygen (O_2) | 0.04% carbon dioxide (CO_2)

It also contains small amounts of other gases, like water vapour and a few noble gases (see p.56). (There's more water vapour than carbon dioxide in the atmosphere.)

The Carbon Dioxide Level is Increasing...

The level of carbon dioxide in the Earth's atmosphere is rising — and it's down to human activities and natural causes. Here are some examples of human activities that affect carbon dioxide levels:

1) Burning fossil fuels to power cars, and to make electricity in power stations, releases lots of carbon dioxide into the atmosphere.

2) Deforestation (chopping down trees) means less carbon dioxide is removed from the atmosphere by photosynthesis.

...Which is Affecting the Earth's Climate

1) Carbon dioxide is what's known as a greenhouse gas. This means it traps energy from the Sun in the Earth's atmosphere. This stops a lot of energy from being lost into space and helps to keep the Earth warm.

This is a bit like what happens in a greenhouse. The Sun shines in, and the glass helps keep some of the energy in.

2) But the level of carbon dioxide (and a few other greenhouse gases) is increasing. The long term trend also shows that the temperature of Earth is increasing. Some scientists believe this is due to the rise in carbon dioxide levels.
3) This increase in the Earth's temperature is called global warming.
4) Global warming is a type of climate change. It could have some serious effects, e.g.
   - Glaciers and ice sheets covering Greenland and Antarctica may melt faster, which could cause sea levels to rise and coastal areas to flood.
   - Rainfall patterns could change, which might make it harder for farmers to grow crops.

The atmosphere and the climate are dead important

The link between carbon dioxide and climate change is a great example of scientists asking questions and developing theories based on what they see in the real world.

Warm-Up and Practice Questions

This section's not too long, so hopefully you'll still remember most of it by this point. The best way to make sure you know all the important stuff covered is to try out some questions and see if you got them right. Well, what are you waiting for? Why not get started on some nice Warm-Up Questions.

Warm-Up Questions

1) Give two things we get from crude oil.
2) Explain why recycling materials means we can burn fewer fossil fuels.
3) What is the role of photosynthesis in the carbon cycle?
4) What gas is most of the Earth's atmosphere made up of?
5) What's a greenhouse gas? Give one example of a greenhouse gas.
6) Why might global warming cause sea levels to rise?

Practice Questions

1 In the UK, local councils collect materials so that they can be recycled.

(a) Give **four** advantages of recycling materials over making them from scratch.

(4 marks)

(b) One local council is designing a leaflet to encourage its residents to recycle more. The leaflet currently includes the three statements below. Say whether each statement is **true** or **false**. If a statement is false, explain why.

1. Recycling could help to prevent further increases in greenhouse gas levels.
2. It is important to recycle aluminium because it uses less energy to recycle aluminium than to mine and extract aluminium from aluminium ore.
3. Recycling saves money because it is completely free.

(2 marks)

2 Mushrooms and earthworms are both types of decomposer. Decomposers are part of the carbon cycle.

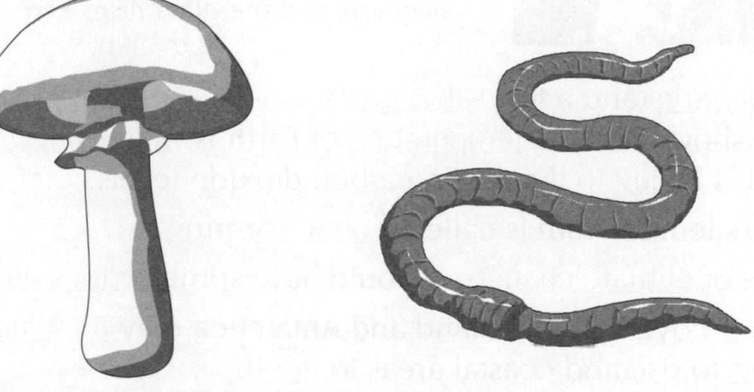

(a) Describe the role of decomposers in the carbon cycle.

(3 marks)

(b) Explain how the role of green plants in the carbon cycle is different to that of decomposers.

(3 marks)

Revision Summary for Section Seven

Well there we are. The end of Section Seven. All you have to do now is learn it all. And yes you've guessed it, here are some lovely questions I prepared earlier. It's no good just idly going through them and managing half-baked answers to one or two that take your fancy. Make sure you can answer them all.

1) The Earth is covered with a thin outer layer of rock. What is this layer called?
2) What is the name of the structure between the outer layer of rock and the Earth's core? Explain how this structure is both a solid and a liquid.
3) Which two metals do we think the Earth's core is made of?
4) What are tectonic plates?
5) How are igneous rocks formed?
6) What determines the size of the crystals in igneous rock?
7) How do sedimentary rocks form?
8) The dead remains of plants and animals can become trapped in sedimentary rocks. What are these remains called?
9) Give two examples of: a) igneous rocks, b) sedimentary rocks, c) metamorphic rocks.
10) Name a mineral present in rocks. Say what elements it contains.
11) Draw out the full diagram of the rock cycle with all the labels.
12) What must happen to sedimentary rocks to turn them into metamorphic rocks?
13) What must happen to metamorphic rocks to turn them into igneous rocks?
14) Name two limited resources we get from the Earth.
15) How is carbon dioxide removed from the air by plants?
16) How does carbon get from the air into your body?
17) How do plants, animals and decomposers all return carbon dioxide to the air?
18) How else is carbon dioxide returned to the air?
19) What percentage of the Earth's atmosphere is: a) nitrogen, b) oxygen, c) carbon dioxide?
20) Name one other gas present in the Earth's atmosphere.
21) Give two human activities that are increasing the level of carbon dioxide in the atmosphere. Say why each one has an effect on the level of CO_2.
22) How does carbon dioxide help to keep the Earth warm?
23) What is global warming? What's causing it?
24) Describe two possible effects of global warming.

Energy Transfer

Everything you do involves energy transfer, which makes these pages pretty important.

Seven Stores of Energy

There are seven stores of energy. Here are some examples of each type:

Gravitational Potential Energy Store

Anything in a gravitational field (i.e. anything that can fall) has energy in its potential energy store — the higher it goes, the more it has.

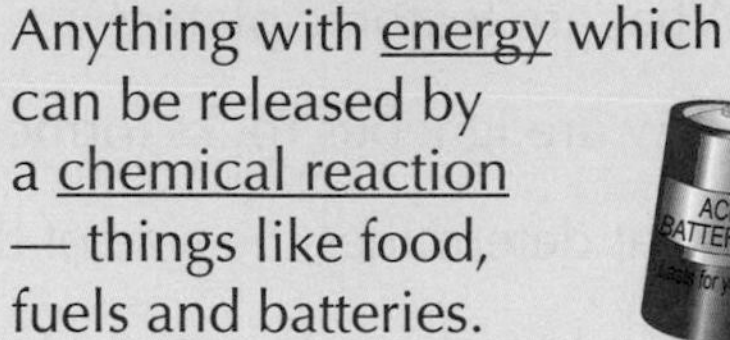

Chemical Energy Store

Anything with energy which can be released by a chemical reaction — things like food, fuels and batteries.

Kinetic (Movement) Energy Store

Anything that moves has energy in its kinetic energy store.

Elastic Potential Energy Store

Anything stretched has energy in its elastic energy store — things like rubber bands, springs, knickers, etc.

Magnetic Energy Store

Two magnets that attract or repel each other have energy in their magnetic energy stores.

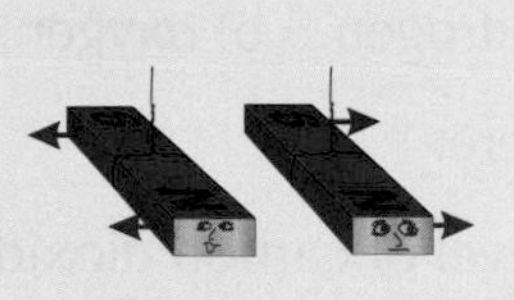

Electrostatic Energy Store

Two electric charges that attract or repel each other have energy in their electrostatic energy stores.

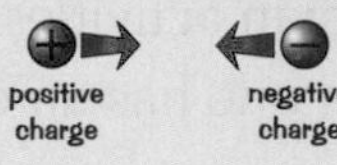

Thermal Energy Store

Everything has some energy in its thermal energy store — the hotter it is, the higher its temperature and the more energy is in its thermal energy store.

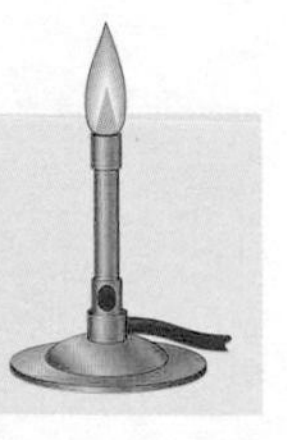

Energy Transfer

Energy Can Be **Transferred** Between Stores

1) Whenever (pretty much) anything happens to an object, energy is transferred from one store to another — the store of energy you transfer to increases and the store of energy you transfer from decreases.
2) There are four main ways you can transfer energy between stores:

Mechanically

When a force makes something move (see page 104). E.g. if an object is pushed, pulled, stretched or squashed.

By **Heating**

When energy is transferred from hotter objects to colder objects (see page 105).

Electrically

When electric charges move around an electric circuit due to a potential difference (see page 151).

By **Light** and **Sound**

When light or sound waves (see Section 10) carry energy from one place to another.

Here are some examples of energy being transferred between stores:

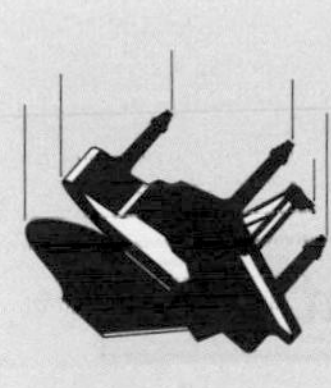

When you drop an object, it moves through a gravitational field. This causes energy to transfer from its gravitational energy store to its kinetic energy store.

When you burn fuel, energy is transferred from the fuel's store of chemical energy to the thermal energy store of the surroundings.

When you switch on this electrical circuit, energy is transferred from a chemical energy store in the battery to a kinetic energy store in the motor. Then as the motor turns, parts of it rub together — this causes some energy to be transferred from the kinetic energy store to a thermal energy store.

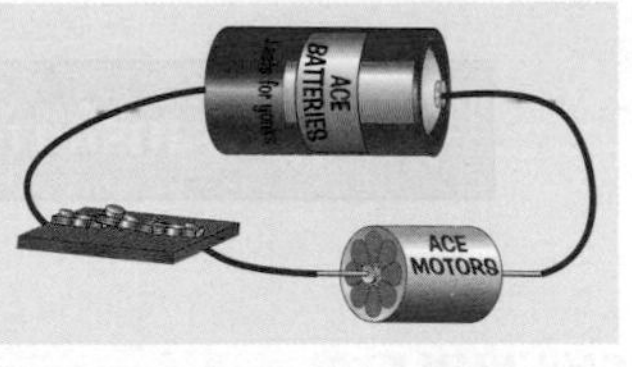

A stretched object, like a spring, has an elastic energy store. When it's released, the energy in the elastic energy store decreases quickly as it is transferred to a kinetic energy store.

Food contains chemical energy stores. When you eat food, it is metabolised (changed during chemical processes inside your body), which releases (transfers) the energy in the food. You can then use the energy for useful things like walking, keeping warm and studying science.

Transferring energy between different stores is really useful

Make sure you know the four ways that energy can be transferred and get your head round all those examples. They're exactly the sort of thing teachers might ask you in Key Stage Three Science.

More Energy Transfer

Energy is stored. And when that energy's transferred via a force, there's a little equation to work out how much energy's transferred.

Energy is Transferred When a Force Moves an Object

When a force moves an object through a distance, energy is transferred.

Energy transferred is the same as work done — see page 130.

1) Whenever something moves, something else is supplying some sort of 'effort' to move it.
2) The thing putting in the effort needs a supply of energy (from fuel or food or electricity etc.).
3) It then transfers energy by moving the object — the supply of energy is transferred to kinetic energy stores.

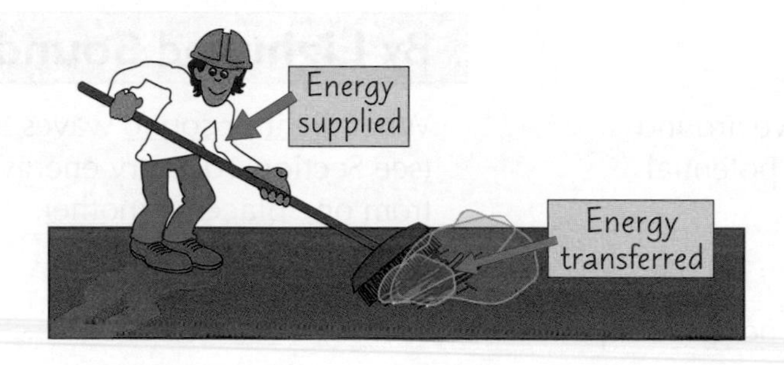

Energy Transferred, Force and Distance are Linked by an Equation

1) To find how much energy has been transferred (in joules), you just multiply the force in N by the distance moved in m.

Energy Transferred (in joules, J) = Force (in newtons, N) × Distance (in metres, m)

EXAMPLE: Some farmers drag an old tractor tyre 5 m over rough ground. They pull with a total force of 340 N. Find the energy transferred.

ANSWER: Energy transferred = force × distance = 340 × 5 = 1700 J

2) So, if a machine transfers a certain amount of energy, the amount of force it can apply and the distance over which it can apply it are linked — if one goes up, the other must come down.
3) So the machine can apply a large force over a small distance, or a small force over a large distance.

Do some work and transfer this into your brain

Work done = energy transferred is a big deal in physics. Make sure you remember it. Without looking at the page, find how far a 2 N force moves an object if it does 20 J of work.

Ans: 10 m.

Energy Transfer by Heating

Energy can be transferred between objects by heating.

Energy is Transferred From Hotter Objects to Cooler Ones

1) When there's a temperature difference between two objects, energy will be transferred from the hotter one to the cooler one (so the hotter object will cool down and the cooler object will heat up).
2) This carries on until the objects reach thermal equilibrium — the point at which they're both the same temperature.

You need to know about two ways in which energy can be transferred between objects by heating:

1) Conduction

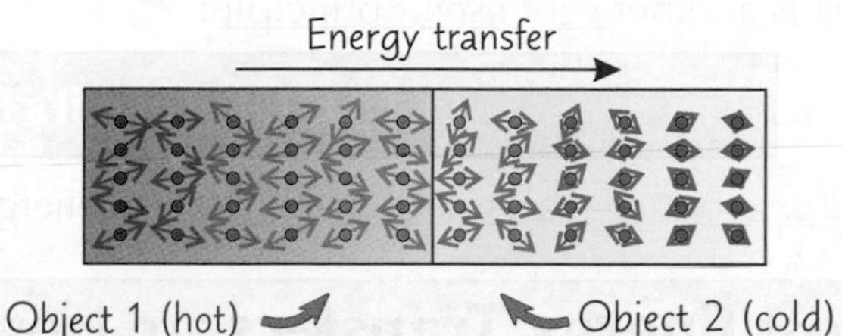

1) When an object is heated, the particles in the object start vibrating more — they gain energy in their kinetic energy stores.
2) Conduction occurs when vibrating particles pass on their extra energy to neighbouring particles.
3) It only happens when particles can bump into each other, so for energy to be transferred from one object to another, they must be touching.
4) Particles in the hotter object vibrate faster than particles in the cooler object. When the particles in the hot object bump into the particles in the cold object, energy is transferred. This means the hot object loses energy and cools down and the cold object gains energy and heats up.

2) Radiation

1) All objects radiate invisible waves that carry energy to the surroundings — the hotter an object is, the more energy it radiates.
2) Radiation isn't transferred by particles, so the objects don't need to be touching.
3) The hotter object (like this hot potato) radiates more energy than the cooler object. The hotter object radiates more energy than it absorbs, so it cools down.
4) The cooler object absorbs some of the radiation from the hot object. It absorbs more energy than it radiates, so it heats up.

Radiation

Hot potato

Cool potato

Insulators Can Slow Down the Rate of Energy Transfer

1) Some materials transfer energy more quickly than others. Objects made from conductors (e.g. metals) will transfer energy more quickly than objects made from insulators (e.g. plastics).
2) Wrapping an object in an insulator will slow down the rate at which it transfers energy to and from surrounding objects. So insulators help keep hot objects hot, and cold objects cold.

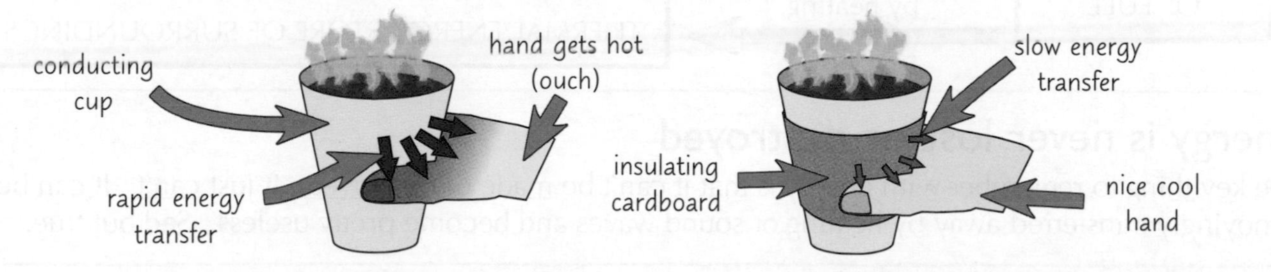

Conservation of Energy

It's not that energy can be transferred, but more that it has to be. Not necessarily in a useful way, mind.

The **Principles** of **Conservation of Energy**

Scientists have only been studying energy for about two or three hundred years so far, and in that short space of time they've already come up with two "Pretty Important Principles" relating to energy. Learn them really well:

THE PRINCIPLE OF CONSERVATION OF ENERGY:
Energy can never be CREATED nor DESTROYED — it's only ever TRANSFERRED from one store to another.

That means energy never simply disappears — it always transfers to another store. This is another very useful principle:

Energy is ONLY USEFUL when it's TRANSFERRED from one store to another.

Think about it — all useful machines use energy from one store and transfer it to another store.

Most **Energy Transfers** are Not Perfect

1) Useful devices are useful because they transfer energy from one store to another.
2) Some energy is always lost in some way, nearly always by heating.
3) As the diagram shows, the energy input will always end up coming out partly as useful energy and partly as wasted energy — but no energy is destroyed:

ENERGY INPUT → USEFUL DEVICE → USEFUL ENERGY OUTPUT; WASTED ENERGY

Total Energy INPUT = The USEFUL Energy + The WASTED Energy

You Can Also Draw **Energy Transfer Diagrams**

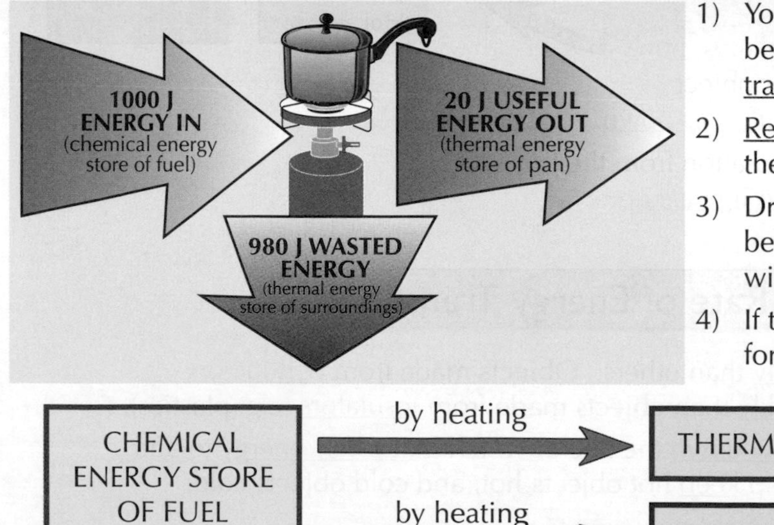

1) You can show how energy moves between stores by drawing an energy transfer diagram (see below).
2) Rectangles are used to represent the different stores.
3) Draw an arrow to show energy being transferred and label it with the method of transfer.
4) If there's more than one transfer, draw an arrow for each one, each going to a different store.

CHEMICAL ENERGY STORE OF FUEL — by heating → THERMAL ENERGY STORE OF PAN

CHEMICAL ENERGY STORE OF FUEL — by heating → THERMAL ENERGY STORE OF SURROUNDINGS

Energy is never lost or destroyed

The key thing to remember with energy is that it can't be made or destroyed. It just can't. It can be annoyingly transferred away by heating or sound waves and become pretty useless. Sad but true.

Warm-Up and Practice Questions

Feeling energised? I hope so, because it's time to see how much of all that energy transfer stuff you've taken in over the last few pages. You know what to do...

Warm-Up Questions

1) Name seven types of energy stores.
2) What units are used to measure energy?
3) Explain what 'radiation' means in terms of energy transfer.
4) Name two ways energy is transferred from a lightbulb.
5) How much energy is transferred when a box is moved 6 metres with a force of 4 newtons?

Practice Questions

1 Ricky was on a caving holiday. He fixed his head torch onto his hat, to help him see when he was inside the dark cave.

(a) When Ricky fixes the head torch onto his hat it is switched off.
Copy and complete the sentence below by selecting the correct option in brackets.

The energy is stored inside the (battery / bulb)
in its (kinetic / chemical) energy store.

(2 marks)

(b) Ricky climbs up the wall of the cave. Copy and complete the flow diagram to show some of the energy transfers that take place when Ricky climbs.

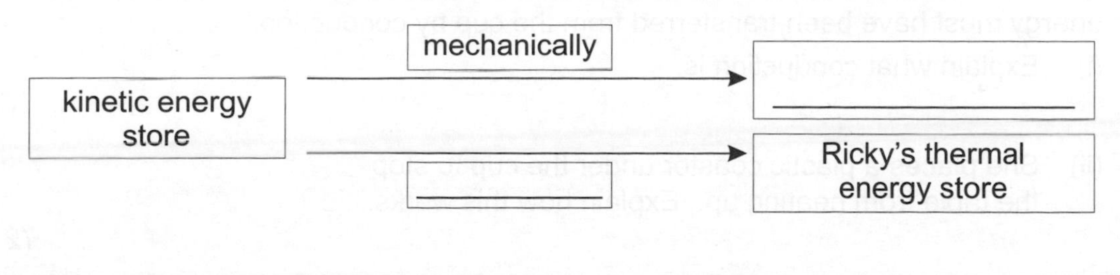

(2 marks)

Practice Questions

2 Copy the sentences below, filling in the gaps to show the energy transfers that have taken place between energy stores.

(a) A wind turbine spins and produces electricity.

Energy in the .... energy store of the turbine is transferred away ....

(b) Coal is burnt in a fireplace to warm a room.

Energy in the .... energy store of the coal is transferred by .... to the .... energy stores of the room.

(c) A skydiver jumps out of a plane.

Energy in the .... energy store of the skydiver is transferred .... to his .... energy store.

(3 marks)

3 Denise is investigating energy transfer by heating. The diagram below shows two identical cups — one containing hot water and one containing an equal amount of cold water.

Hot water

Cold water

(a) Denise places her hand near the cup of hot water and notices that her hand feels warmer.

(i) Explain what causes her hand to feel warmer.

(1 mark)

(ii) Explain why this doesn't happen when she places her hand near the cold cup.

(1 mark)

(b) Denise notices that the table around the hot cup feels warm, and decides energy must have been transferred from the cup by conduction.

(i) Explain what conduction is.

(1 mark)

(ii) She places a plastic coaster under the cup to stop the table from heating up. Explain how this works.

(2 marks)

Energy Resources

The Sun's a useful little critter. It provides us with oodles of energy and asks for nothing in return.

The **Sun** is the **Source** of Our **Energy Resources**

Most of the energy around us originates from the Sun. The Sun's energy is really useful for supplying our energy demands. Often the Sun's energy is transferred to different stores before we use it.

Learn These **Six** Energy Transfer Chains

1. Sun's Energy ⟶ **Coal**, **Oil**, and **Gas** (Fossil Fuels)

Sun ➔ light ➔ photosynthesis ➔ dead plants/animals ➔ FOSSIL FUELS

2. Sun's Energy ⟶ **Biomass** (e.g. Wood)

Sun ➔ light ➔ plants ➔ photosynthesis ➔ BIOMASS (wood)

3. Sun's Energy ⟶ **Food**

Sun ➔ light ➔ plants ➔ photosynthesis ➔ BIOMASS (food)

4. Sun's Energy ⟶ **Wind Power**

Sun ➔ heats atmosphere ➔ air circulates ➔ causes WINDS

5. Sun's Energy ⟶ **Wave Power**

Sun ➔ heats atmosphere ➔ causes WINDS ➔ causes WAVES

We can use the energy in the kinetic energy stores of the wind and waves to turn turbines and generators, giving us electricity (see page 110).

6. Sun's Energy ⟶ **Solar Cells**

Sun ➔ light hits solar cells ➔ generates ELECTRICITY

Generating Electricity

We can use the energy we get from the Sun to generate electricity, in lots of different ways...

There Are Different Ways of Generating Electricity

1) There are a variety of different fuels that people use in their homes, e.g. coal is used for fires, gas is used for cookers, etc. But most homes these days rely on electricity for most of their energy needs.
2) We can use energy resources (see previous page) to generate electricity.
3) At the moment we generate most of our electricity by burning fossil fuels.
4) Most ways of generating electricity use an energy resource to turn a turbine and a generator. Energy is transferred between different stores before being transferred away electrically.
5) Energy resources that we use to generate electricity can be split into two groups — non-renewable and renewable.

Boiler
Turbine
Generator
Fuel
Grid
Chemical energy store → Thermal energy store → Kinetic energy store → Kinetic energy store → Transferred away electrically

Non-renewable Energy Resources Will Run Out

1) Fossil fuels took millions of years to come about — and only take minutes to burn.
2) Once they've been taken from the Earth — that's it, they're gone, (unless you're gonna wait around a few more million years for more to be made).
3) There'll come a time when we can't find any more and then we could have a problem.
4) We need to reduce the amount of fossil fuels we use, so they won't run out as quickly. The answer is:

   i) Save energy (e.g. turn lights off, drive cars with more fuel-efficient engines).
   ii) Recycle more (see page 97).
   iii) Use more renewable energy resources (see below).

Renewable Energy Resources Won't Run Out

As long as the Sun still shines...

1) The WIND will always blow — and turn turbines to generate electricity.
2) PLANTS will always grow — which can be burnt to generate electricity.
3) WAVES will always be made — and drive generators to make electricity.
4) SOLAR cells will always work — and use light to make electricity.

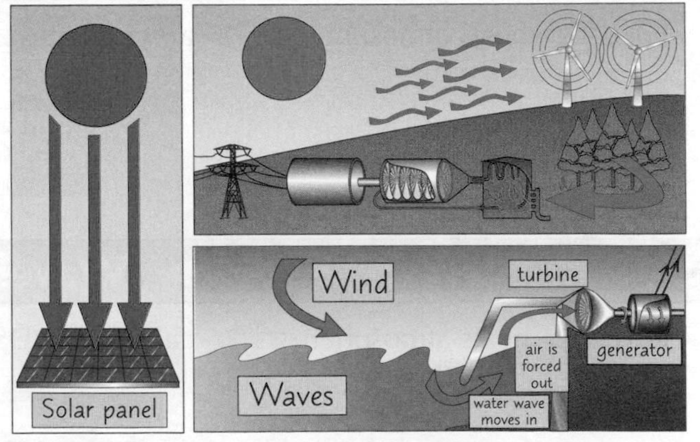

One of the drawbacks of renewable energy is that it's not always available like non-renewable resources. For example, there will always be wind, but it isn't always blowing with the same strength, solar cells are only useful during the day etc.

Think on this — it'll affect all your generation

A common mistake is to call renewable energy resources "re-usable". They can be renewed like trees will grow again if replanted, etc. But once a tree is burnt you can't re-use it.

The Cost of Electricity

To work out the cost of electricity, you first need to know how to calculate the energy transferred.

You Can Calculate the Energy an Appliance Transfers

1) Anything that needs electricity to work is an electrical appliance.
2) All electrical appliances transfer energy electrically to stores of energy. Energy can be measured in joules (J), kilojoules (kJ) or kilowatt-hours (kWh).
3) Power tells you how fast something transfers energy. It's usually measured in watts (W) or kilowatts (kW). 1 kW = 1000 W.
4) The total amount of energy transferred depends on the amount of time something's switched on for, and its power. If you know the power in watts and the time in seconds, you can calculate energy transferred using this equation:

ENERGY TRANSFERRED (J) = POWER (W) × TIME (s)

EXAMPLE: A 2000 watt kettle boils in 150 seconds. Calculate the energy transferred.

ANSWER: Energy transferred (J) = power (W) × time (s)
Energy transferred = 2000 × 150 = 300 000 J

5) If you know the power in kilowatts and the time in hours, you can use this equation:

ENERGY TRANSFERRED (kWh) = POWER (kW) × TIME (h)

This gives the energy transferred in kilowatt-hours — the number of kilowatts used each hour.

A kilowatt is 1000 watts

EXAMPLE: Andrea uses a 3 kilowatt heater for 4 hours to heat her house on a cold day. Calculate the energy transferred.

ANSWER: Energy transferred (kWh) = power (kW) × time (h)
Energy transferred = 3 × 4 = 12 kWh

Calculate energy transferred using power and time

So you've got two versions of the same formula up there — if you look carefully, you'll see that only the units have changed. Make sure you always check the units of any data you have, and if necessary convert it to the right units before you do any calculations.

The Cost of Electricity

Electricity isn't free you know — ask your mum and dad. But at least the cost is pretty easy to calculate.

Electricity Meters Record How Much Electricity is Used

Electricity meters record the amount of energy transferred in kWh. You can use them to work out the energy transferred over different periods of time, e.g. at day and at night:

EXAMPLE: Ganesh wants to find out how much electricity he uses during the day compared to during the night. He writes down his meter reading at three different times during a 24-hour period:

6pm = 44281.25 kWh
6am = 44284.76 kWh
6pm = 44296.12 kWh

Does he use more electricity during the day or during the night?

ANSWER: Energy from 6pm to 6am (i.e. during the night) = 44284.76 – 44281.25 = 3.51 kWh
Energy from 6am to 6pm (i.e. during the day) = 44296.12 – 44284.76 = 11.36 kWh
So he uses more electricity during the day.

Calculating the Cost of Electricity

Domestic fuel bills charge by the kilowatt-hour.
You can calculate what your electricity bill should be with this handy little formula:

COST = Energy transferred (kWh) × PRICE per kWh

Cost = kWh × Price

EXAMPLE: Electricity costs 16p per kWh. At the start of last month, Jo's electricity meter reading was 42729.66 kWh. At the end of the month it was 43044.91 kWh. Calculate the cost of her electricity bill last month.

ANSWER: Energy transferred = 43044.91 – 42729.66 = 315.25 kWh
Cost = Energy transferred × Price = 315.25 × 16 = 5044p = £50.44

Many homes use gas as a fuel, e.g. for gas central heating, gas cookers etc.
Your gas bill is calculated using the energy used in kWh, just like your electricity bill.

Learn how to calculate the cost of electricity

Ever wondered what those little numbers on your electricity meter tell you? It's good to check the power company's charged you the right amount every now and then — and now you can do just that with a nice little formula. Make sure you know how to use it — it's dead useful.

Comparing Power Ratings and Energy Values

If you want to know how much energy an appliance uses you can work it out using its power rating. And if you want to know how much energy is in your food, just look on the label.

Power Ratings of Appliances

1) The power rating of an appliance is the energy that it uses per second when it's operating at its recommended maximum power (i.e. when it's plugged into the mains).
2) You can work out the energy transferred by an appliance in a certain time if you know its power rating. To do this you need to use the equations on page 111.

The Energy Transferred Depends on the Power Rating

The higher the power rating of an appliance, the more energy it transfers in a given amount of time. You can compare how much energy is transferred by appliances with different power ratings.

EXAMPLE: How much energy is transferred by a 1.5 kW electric heater compared to a 4 kW electric heater, when they're both left on for 1.5 hours?

ANSWER: Energy transferred (kWh) = power rating (kW) × time (h).
Energy transferred by the 1.5 kW heater = 1.5 × 1.5 = 2.25 kWh.
Energy transferred by the 4 kW heater = 4 × 1.5 = 6 kWh.

So the 4 kW heater transfers (6 – 2.25) 3.75 kWh more energy than the 1.5 kW heater in 1.5 hours.

Remember, transferring energy costs money. So an appliance with a higher power rating will cost more to run over a set period of time than an appliance with a lower power rating.

Food Labels Tell You How Much Energy is in Food

1) All the food we eat contains energy — it's important to make sure you're taking in the right amount of energy each day (page 9).
2) The energy in food is measured in kilojoules (kJ).
3) You can compare the amount of energy in different foods by looking at their label.

You may also see a value for kcals on a food label — this is just another unit that energy can be measured in.

Warm-Up and Practice Questions

You're nearly done in this section now. Take time to make sure you've really understood everything so far — these questions will help you discover any gaps in your knowledge.

Warm-Up Questions

1) How is energy from the Sun transferred to the chemical energy stores of biomass?
2) What does an electricity meter in your home measure?
3) What is the equation you would use to calculate the cost of electricity?
4) How much energy is transferred by a 60 W lightbulb left on for 30 minutes?
5) Where does most of our energy originate from?
6) Suggest two ways in which we can use less energy from non-renewable resources.

Practice Questions

1 Natural gas is an important fossil fuel used in power stations and in the home.

(a) Natural gas is described as a 'non-renewable energy resource'. Explain what this term means.

(1 mark)

(b) Give **two** examples of renewable energy resources.

(2 marks)

(c) Explain why it is important to use more renewable energy resources.

(1 mark)

2 Last week, Andrew used his 3 kW kettle for a total of 1.75 hours.

(a) How much energy in kWh did Andrew's kettle transfer last week?

(1 mark)

(b) Electricity costs 17p per kWh. Work out the cost of the electricity used by Andrew's kettle last week.

(1 mark)

3 Anne is testing a new energy-saving lightbulb with a low power rating.

(a) What does 'power rating' mean?

(1 mark)

(b) The new energy-saving lightbulb is a 10 W bulb.

(i) How much energy does the lightbulb transfer when it's used for one hour?

(1 mark)

(ii) Anne usually uses a 60 W bulb. How much more energy is transferred by the 60 W bulb compared to the 10 W bulb when they're both left on for an hour?

(2 marks)

Physical Changes

A change of temperature can change a substance physically.

Physical Changes Don't Involve a Change in Mass

1) A substance can either be a solid, a liquid or a gas. These are called states of matter. When a substance changes between these physical states, its mass doesn't change.
2) Physical changes are different to chemical changes because there's no actual reaction taking place and no new substances are made. The particles stay the same, they just have a different arrangement and amount of energy.

There are several different processes that can change the physical state of a substance: melting, freezing, condensing, evaporating, dissolving and sublimation.

Melting, Evaporating, Condensing, Freezing

1) If you melt a certain amount of ice, you get the same amount of water.
2) If you boil the water so it evaporates, you get the same amount of steam.
3) It's the same in the other direction — if the steam condenses, you get the same amount of water.
4) And if the water freezes, you get the same amount of ice.

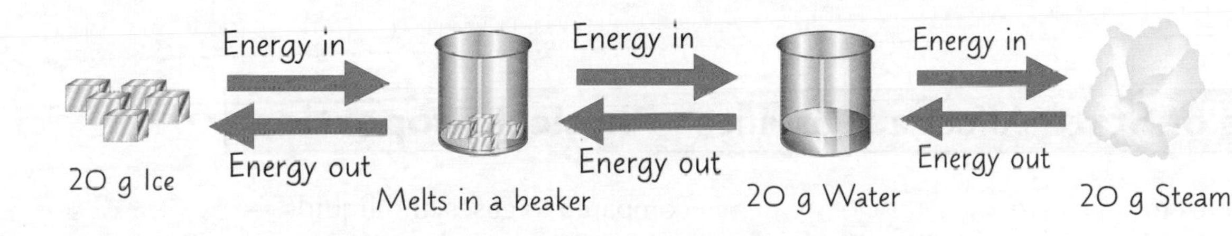

Sublimation is a Change from a Solid to a Gas

1) Some substances, such as carbon dioxide, can go straight from being a solid to being a gas.
2) This is called sublimation.
3) When this happens, the mass of gas is (you guessed it) the same as the mass of the solid.

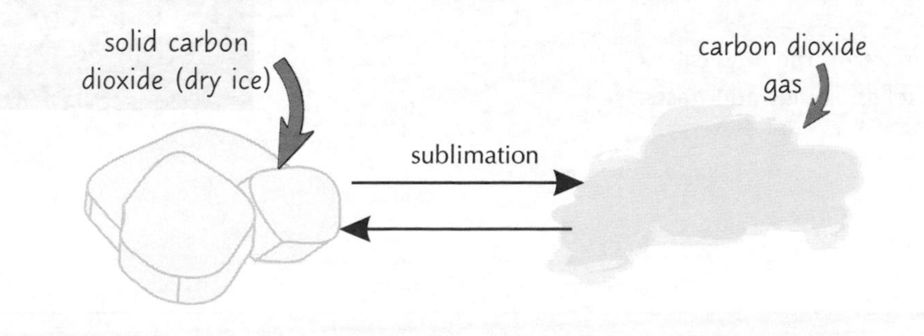

Remember, changes of state don't change the mass

It may not seem like it, but boiling off a beaker of water makes a cloud of water vapour with the same mass as the initial water. The same happens with freezing — 100 g of water makes 100 g of ice.

Physical Changes

You saw some physical changes on the last page — there are more here that you need to learn too.

Dissolving — There's No Change in Mass

1) When a solid substance dissolves to form a solution (page 61), there's no change in mass. The amount of substance after dissolving is the same as before, it's just in a different form.
2) Dissolving is reversible — if you evaporate all the solvent, you'll be left with the same amount of solid as before it dissolved.

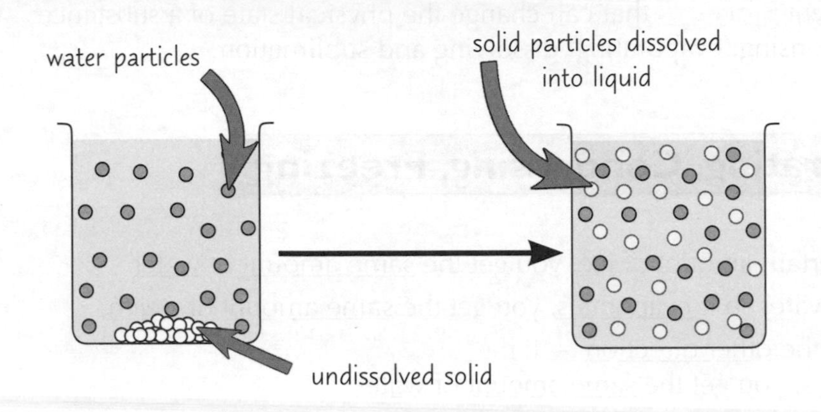

Changes of State Affect a Substance's Physical Properties

1) The particles in solids are packed together tightly compared to gases and liquids — so they're usually more dense. They're also difficult to compress and can't flow.
2) The particles in liquids and gases are free to move around each other, so they can flow.
3) When you heat a substance, the particles move around more and move further apart, causing it to change from a solid, to a liquid, to a gas. The substance expands and becomes less dense.
4) Ice is a funny one though — when it melts (to become water), the particles actually come closer together and its density increases. It's why icebergs float on water.

See p.49 for more on the physical properties of solids, liquids and gases.

Changes of state change a substance's physical properties

Remember, physical changes have no effect on the chemical structure of the substance — they just cause it to change state between solid, liquid and gas. Ice is really just water, but with the particles in a more structured formation. If it melts, it goes back to being water without losing any mass at all. And if you dissolve salt in water then boil the water off, you get the same mass of salt back.

Movement of Particles

Brownian motion isn't to do with colours — Brown was just the name of the guy who first noticed it.

Brownian Motion is the Random Movement of Particles

Atoms and molecules are both types of particle.

1) In 1827, a scientist called Robert Brown noticed that tiny pollen particles moved with a zigzag, random motion in water.
2) This type of movement of any particle suspended ('floating') within a liquid or gas is known as Brownian motion.
3) Large, heavy particles (e.g. smoke) can be moved with Brownian motion by smaller, lighter particles (e.g. air) travelling at high speeds — which is why smoke particles in air appear to move around randomly when you observe them in the lab.

Diffusion is Caused by the Random Motion of Particles

1) The particles in a liquid or gas move around at random.
2) Particles eventually bump and jiggle their way from an area of high concentration to an area of low concentration. They constantly bump into each other, until they're evenly spread out across the substance. This is called diffusion.

High concentration of purple particles

Low concentration of purple particles

The particles move about randomly until...

...there's an even concentration of purple particles on each side.

Movement of Particles Increases with Temperature

1) An increase in temperature causes particles to move around more — their speed increases. This means the spaces between the particles get bigger so they take up more space, which causes the material to expand.
2) If you heat a gas or liquid inside a container, the pressure inside the container increases as the particles bump into the sides more often and more quickly.
3) The particles don't get bigger — they just need more space to move around in.

EXAMPLE:
When the liquid in this flask is heated its volume expands as the particles move apart with their extra energy.

So the liquid moves up the thin tube — this is how a thermometer works — dead useful...

Particles move randomly — this is known as Brownian motion

You need to know how particles move in liquids and gases. Once you've done that, you can use it to explain why diffusion happens and how particles gradually spread out. Don't forget that temperature affects how fast particles move (see page 50) — they move a lot faster in hot liquids or gases.

Warm-Up and Practice Questions

It's the end of another section. Energy can be a bit of a tricky topic to understand, so take your time to work through these questions until you're sure.

Warm-Up Questions

1) If you boil 100 g of water until there is no liquid left, what mass of water vapour will you get?
2) What is sublimation?
3) What causes smoke particles to move around at random in air?
4) Give two differences between the physical properties of a gas and a solid of the same substance.

Practice Questions

1 Sally dissolves 36 g of salt in 100 g of water at room temperature.

(a) What is the mass of the salt solution?

(1 mark)

(b) Suggest how Sally could get the salt back out of the solution.

(1 mark)

2 Particles in a gas move from areas of high concentration to low concentration.

(a) What name is given to this motion?

(1 mark)

(b) Explain how this motion causes particles to become evenly spread out throughout a gas.

(2 marks)

3 A balloon is filled with helium gas. It floats by a light bulb and heats up.

(a) What effect will this have on the motion of the helium particles inside the balloon?

(1 mark)

(b) What effect will this have on the size of the balloon? Explain your answer.

(2 marks)

4 Solids and liquids display different physical properties.

(a) (i) Describe the general difference in density between a solid and a liquid of the same substance.

(1 mark)

(ii) Give one substance that your answer to part (i) does not apply to.

(1 mark)

(b) Describe one difference between the physical properties of solids and liquids other than their density.

(1 mark)

Revision Summary for Section Eight

Ah, the section summary — on the home stretch at last. In this section, you got to go on a voyage of discovery into the weird and wonderful world of all things energy and matter. All that's left for you now is to work through the exciting questions below and claim your free ice cream at the start of the next section. OK, the ice cream is a lie, but you won't regret taking the time to work through these questions if you want to be super-amazing at science, trust me. Take your time with them if you like, and maybe have the odd cheeky peek back at the appropriate page if you're stuck — I won't tell anyone, honest.

1) Name the seven types of energy stores and the four ways of transferring energy between them.
2) Give one example of energy being transferred electrically to one or more energy stores.
3) Describe the energy transfers for an object that is falling.
4) When does an object have energy stored in its chemical energy store?
5) Replace the gaps in the sentence below with the correct words.
When a .... moves an object through a distance, ... is transferred.
6)* A crane applies a force of 2000 N to lift a small elephant 10 m.
How much energy does it transfer?
7) What does thermal equilibrium mean?
8) Name two ways in which energy can be transferred between two objects by heating.
9) Describe how energy is transferred when two objects at different temperatures are touching.
10) How does adding an insulator to an object affect the rate of energy transfer?
11) What is the Principle of Conservation of Energy?
12) Why is it important that devices transfer energy from one store to another?
13) Why are most energy transfers NOT perfect?
14) How is wasted energy usually transferred?
15) How does the Sun's energy get stored in fossil fuels?
16) Other than fossils fuels, give two energy resources created using the Sun's energy.
17) What are non-renewable energy resources?
18) What are renewable energy resources? Why will they never run out?
19)*Calculate the energy transferred by a 1.5 kW remote-control car used for half an hour.
20) What unit is household electricity measured in?
21)*Electricity costs 15p per kWh. Calculate the cost of an electricity bill for 298.2 kWh.
22) What does the power rating of an appliance tell you?
23)*Which will transfer more energy — a 200 W device left on for 1 hour, or a 300 W device left on for 1 hour?
24) What unit is the energy in food usually measured in?
25) Give an example of a physical change of state.
26)*50 g of iron is melted. How much liquid iron would be produced?
27) What's meant by Brownian motion?
28) Explain why gases expand when they're heated.

*Answers on page 196.

Speed

Yes, it's a page on speed. Make sure you can do these calculations — don't zoom through.

Speed is How Fast You're Going

1) Speed is a measure of how far you travel in a set amount of time.
2) The formula triangle is definitely the best way to do speed calculations.

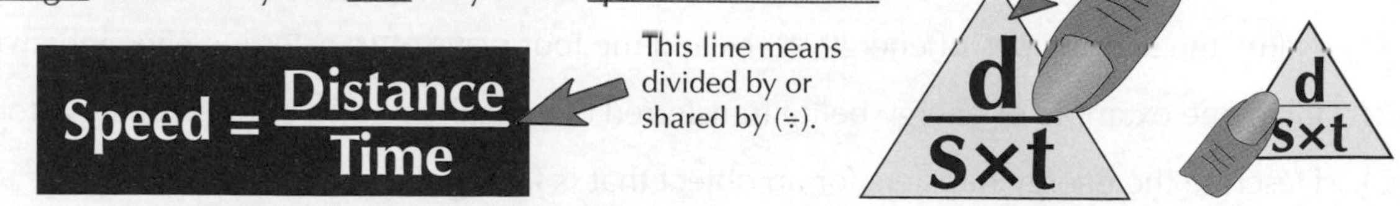

3) Use the word SIDOT to help you remember the formula:

SIDOT — Speed Is Distance Over Time.

Always use UNITS.

4) There are three common units for speed. You should realise that they're all kind of the same, i.e. distance unit per time unit.

- metres per second — m/s
- miles per hour — mph or miles/h
- kilometres per hour — km/h

Work Out Speed Using Distance and Time

To work out speed you need to know the distance travelled and the time taken.

Example 1: A sheep moves down a farmer's track. It takes exactly 5 seconds to move between two fence posts, 10 metres apart. What is the sheep's speed?

Answer

Step 1) Write down what you know:
distance, d = 10 m   time, t = 5 s

Step 2) We want to find speed, s
from the formula triangle: s = d / t
Speed = Distance ÷ Time = 10 ÷ 5 = 2 m/s

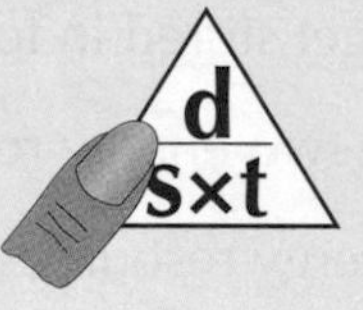

Put your finger over "s" in the formula triangle — which leaves d/t (i.e. d÷t).

Example 2: A van drives down a road and travels 15 miles in 30 minutes. What's its speed?

Answer

Step 1) Write down what you know:
distance, d = 15 miles   time, t = 30 minutes = 0.5 of an hour.

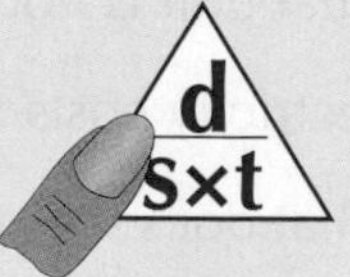

Step 2) We want to find speed, s, from the formula triangle: s=d/t
Speed = Distance ÷ Time = 15 ÷ 0.5 = 30 miles/hour (mph)
For the answer to be in miles per hour you need the distance in miles and the time in hours. So the 30 mins had to become 0.5 hrs.

Ahh good ol' formula triangles

They're brilliant. Use them whenever you can. Remember, the thing at the top of the triangle equals the two things on the bottom, multiplied together. This is true for any formula triangle.

Distance-Time Graphs

Distance-time graphs can seem confusing — read this page extra-carefully.

Distance-Time Graphs Tell You About an Object's **Motion**

A distance-time graph shows the distance travelled by an object over time.

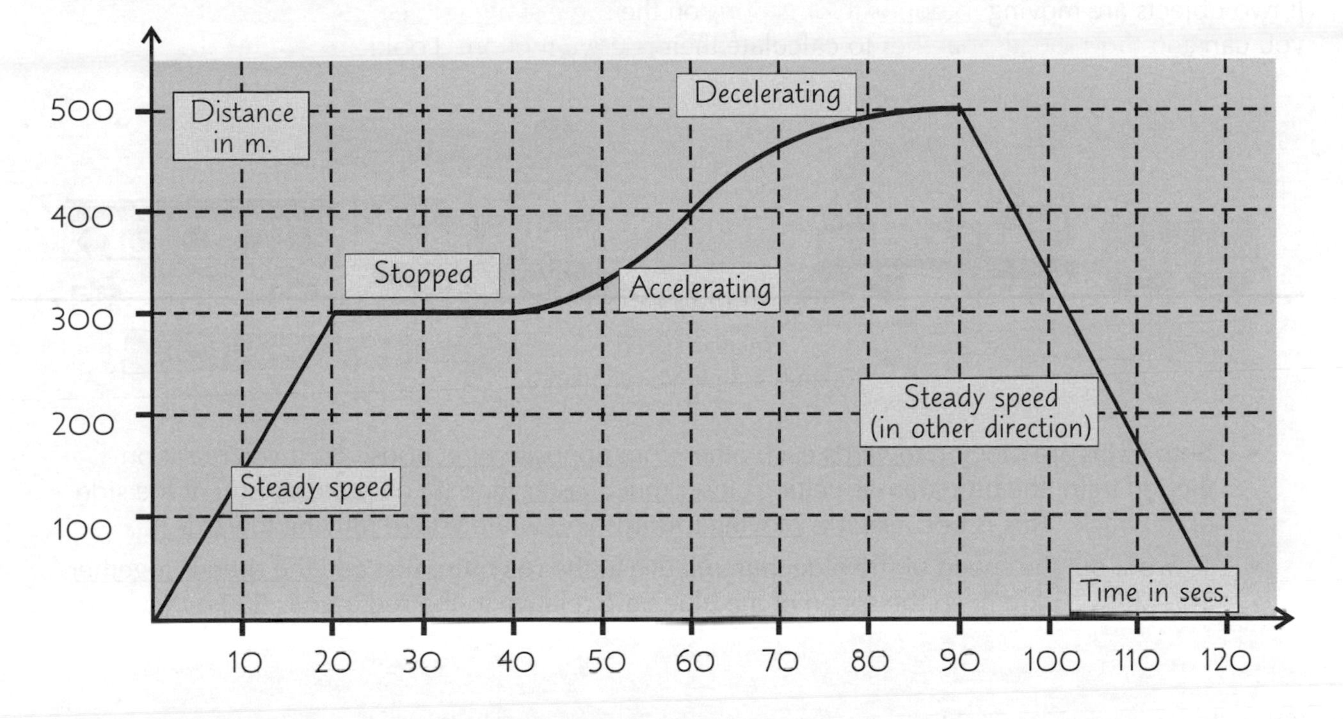

1) The slope of the line (gradient) shows the speed at which the object is moving.

2) The steeper the graph, the faster the object is going.

3) Flat sections are where it's stopped.

4) Downhill sections mean it's moving back toward its starting point.

5) Curves represent a changing speed.

6) A steepening curve means the object is speeding up (accelerating).

7) A curve levelling off means the object is slowing down (decelerating).

The slope of the line is really important

If the line on a distance-time graph is flat, the object's not moving. If it's straight but not flat, the object's moving at a constant speed. But if the line is curved, the object is speeding up (accelerating) or slowing down (decelerating). Get that learnt and these graphs will make all sorts of sense.

Relative Motion

If two objects are moving towards or away from each other, it's useful to calculate their relative motion.

Relative Motion — Two Objects Passing Each Other

Relative motion is useful if you want to know the speed of something when you are moving too.

1) If two objects are moving in opposite directions on the same straight line you can add their speeds together to calculate their relative motion. Look:

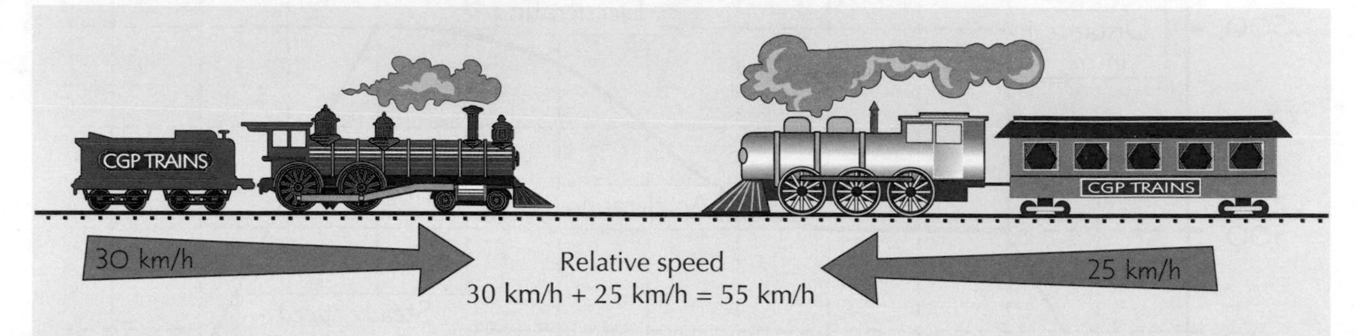

- Both trains are moving towards each other from opposite directions. So if you're sat on the red train, the blue train is getting closer much faster than if you were sat still at the side of the track. This is because it is moving towards you while you're moving towards it.
- To work out the speed of the blue train relative to the red train, just add the speeds together. 30 + 25 = 55 km/h, so the speed of the blue train relative to the red train is 55 km/h.

2) If the objects are moving in the same direction on the same straight line you can subtract their speeds to calculate their relative motion.

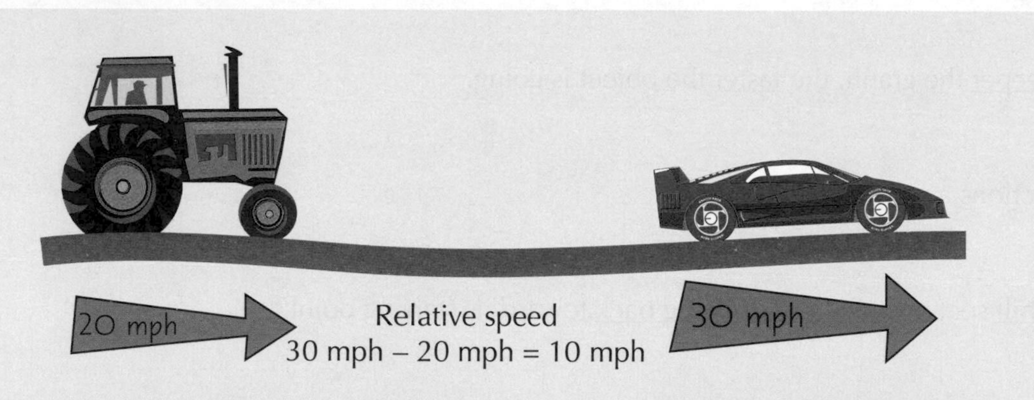

- The car is moving in the same direction as the tractor but at a faster speed.
- If you're in the car, you're getting further away from the tractor more slowly than if it wasn't moving (since it's moving towards you while you're moving away from it)
- To work out the speed of the car relative to the tractor, subtract the speeds. 30 – 20 = 10 mph — the car gets 10 miles further away from the tractor every hour.

Relative motion — for when two things are moving

Imagine two race cars on a track. If you want to know how fast one race car is catching up to the other, you'll need to use relative motion. It's a dead handy tool to use when you've got two objects in motion. Make sure you can calculate it, I have a sneaky feeling you'll be asked about it later.

Forces and Movement

Well, I can't force you to read this page — but if I were you, I'd push on with it...

Forces are Nearly Always Pushes and Pulls

1) Forces are pushes or pulls that occur when two objects interact.
2) Forces can't be seen, but the effects of a force can be seen.
3) Forces are measured in newtons — N.
4) They usually act in pairs.
5) They always act in a certain direction.
6) A newton meter is used to measure forces.

Objects don't need to touch to interact. The gravitational pull between planets (p.162), forces between magnets (p.158) and forces due to static electricity (p.157) are all non-contact forces.

Forces Can Make Objects Do Five Things

1. Speed Up or Start Moving	Like kicking a football. To start something moving, a push force must be larger than resisting forces like friction (see next page).	3. Change Direction	Like hitting a ball with a bat or gravity causing footballs to come back down to Earth.
		4. Turn	Like turning a spanner.
2. Slow Down or Stop Moving	Like drag or air resistance (see p.125).	5. Change Shape	Like stretching and compressing (see p.130), bending and twisting.

Learn These Two Important Statements:

Balanced Forces produce No Change in Movement

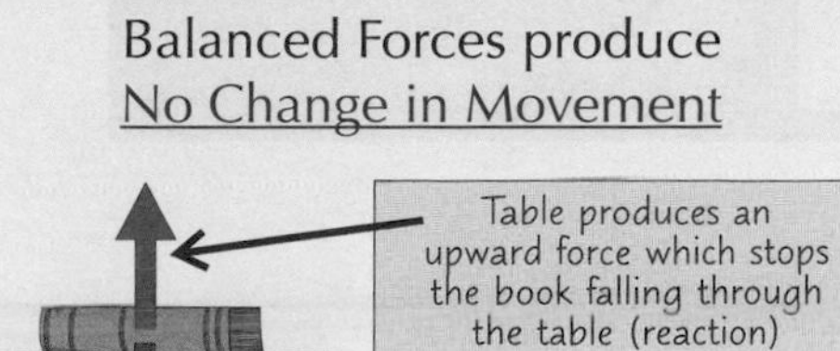

Unbalanced Forces Change the Speed and/or Direction of Moving Objects

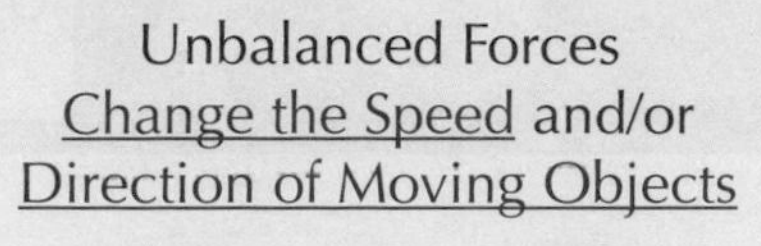

These are force diagrams. See page 128 for more.

Friction

Friction stops you falling over, but slows you down when you try to move. It's a mixed bag.

Friction Tries to Stop Objects Sliding Past Each Other

Friction is a force that always acts in the opposite direction to movement.
It's the force you need to overcome when pushing an object out of the way.

The Good Points of Friction — It Allows Things to Start and Stop

1) Friction allows the tyres on a bike to grip the road surface — without this grip you couldn't make the bike move forward and you wouldn't be able to stop it either. It'd be like riding on ice.
2) Friction also acts at the brakes where they rub on the rim of the wheel or on the brake disc.
3) Friction also lets you grip the bike — important if you want to ride it without slipping off.

The Bad Points of Friction — It Slows You Down

1) Friction always wastes energy — friction between the moving parts of a bike warms up the gears and bearings — a waste of energy.
2) Friction limits top speed. The air resistance (a kind of friction, see next page) takes a lot of your energy and limits your maximum speed.

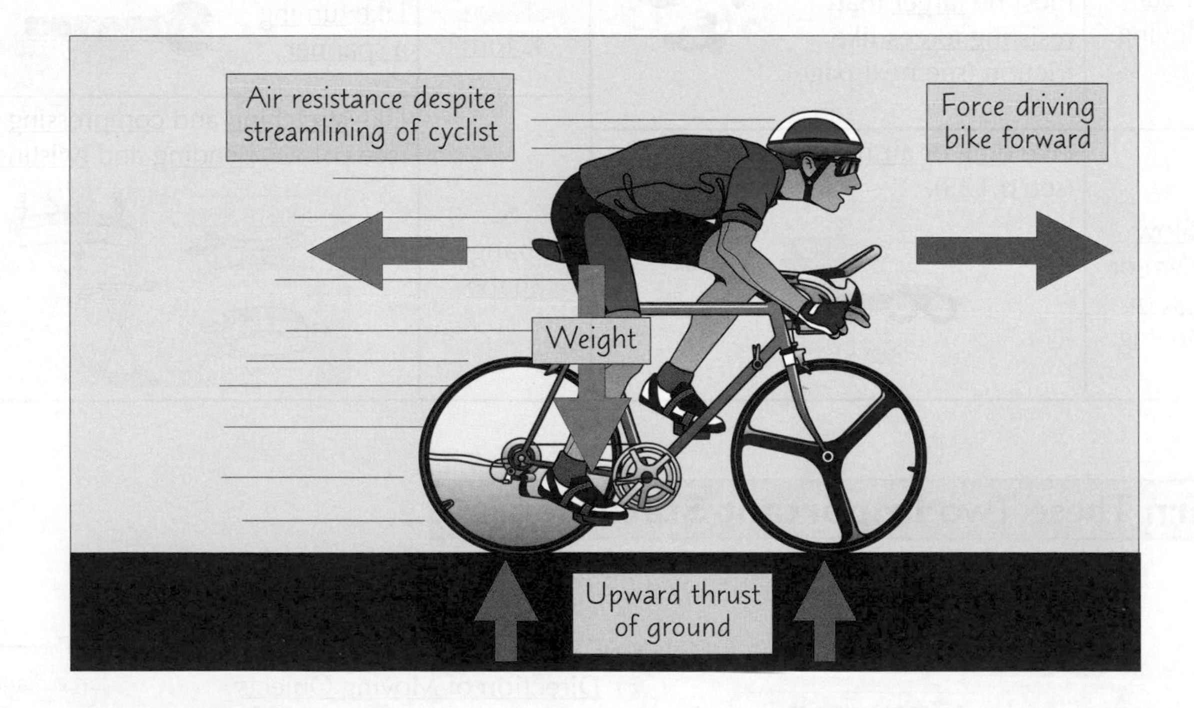

Forces mean lots of arrows to draw

REVISION TASK

This example of the bike is a classic one. But the principles are the same whatever example you might come across. Draw a man on a sledge being pulled by a group of dogs. Label all of the forces acting on the sledge, including the directions in which they each act.

Air and Water Resistance

Air and **Water Resistance** Slow Down **Moving Objects**

1) Air and water resistance (or "drag") push against objects which are moving through the air or water.
2) These are kinds of frictional force because they try to slow objects down.
3) If things need to go fast, then they have to be made very streamlined — which just means they can slip through the air or water without too much resistance. A good example is a sports car.

Air Resistance **Slows** Parachutists Down

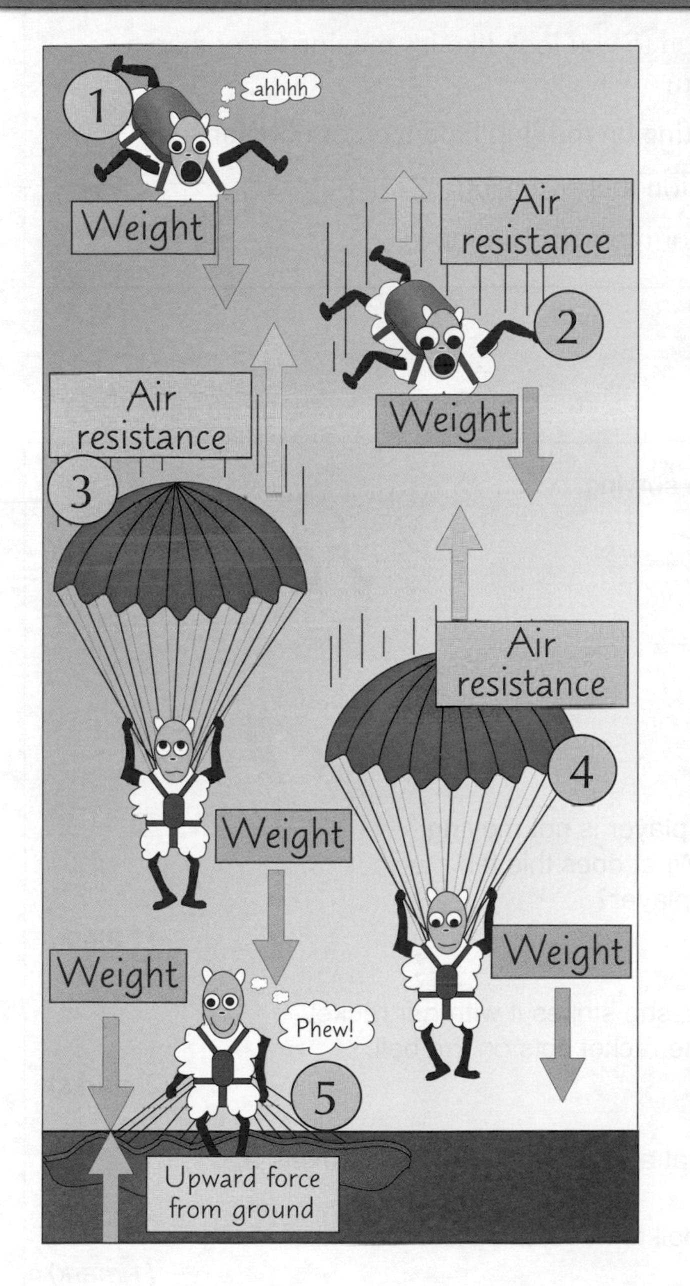

1) **Gains** Speed

At the start, the sheep only has the force of its weight (i.e. gravity) pulling it down — so it starts to move faster.

2) **Still Gaining** Speed

As it moves faster, the opposing force of air resistance gets more and more.

3) **Losing** Speed

When the parachute opens air resistance increases enormously — because there's a much larger area trying to cut through the air. The sheep loses speed and slows down gratefully.

4) **Steady** Speed

Very quickly the air resistance becomes equal to the weight — the two forces are balanced. The overall force is zero, so the sheep now moves at a steady speed.

5) **No** Speed

Once safely on the ground, the sheep's weight acting downwards is balanced by an equal upward force from the ground.

Parachuting sheep — don't try this at home

Anything moving through air or water feels a resistance trying to slow it down. Some objects like boats and cars are streamlined to minimise resistance. You'll also need to understand that resistance makes objects fall eventually at a steady speed by acting in the opposite direction to gravity.

Warm-Up and Practice Questions

Here's another set of Warm-Up and Practice Questions. Take time to give these questions a good go — they'll help you to find out which pages you've understood really well, and which pages could do with a bit more attention. Go on, you know you want to...

Warm-Up Questions

1) How is speed calculated? What are the units of speed?
2) A girl cycles for 30 seconds at 5 m/s. How far has she cycled?
3) How would you show acceleration on a distance-time graph?
4) Why does a car driving in the opposite direction to you look like it's moving faster than a car driving at the same speed in the same direction?
5) A ship's speed is increasing. Are the forces acting on the ship balanced or unbalanced?
6) Which force always acts in the opposite direction to movement?
7) Why does a parachutist fall at a steady speed for most of his jump?

Practice Questions

1 A tennis player tosses her ball in the air before serving.

(a) As she waits for the ball to fall, the tennis player is not moving, although there are forces acting on her. What does this tell you about the forces acting on the tennis player?

(1 mark)

(b) As the ball begins to fall back towards her, she strikes it with her racket. Suggest **two** ways in which the force of the racket acts on the ball.

(2 marks)

(c) The ball leaves the tennis player's racket at a speed of 50 m/s and travels a distance of 20 m before bouncing.

(i) Calculate how long it takes the ball to travel this distance.

(1 mark)

(ii) The actual time taken for the ball to travel this distance is slightly longer than you have calculated. Suggest a reason why.

(1 mark)

Practice Questions

2 Trevor plays with his model train for 80 seconds. He plots the train's distance from its start point over this time period on the graph below.

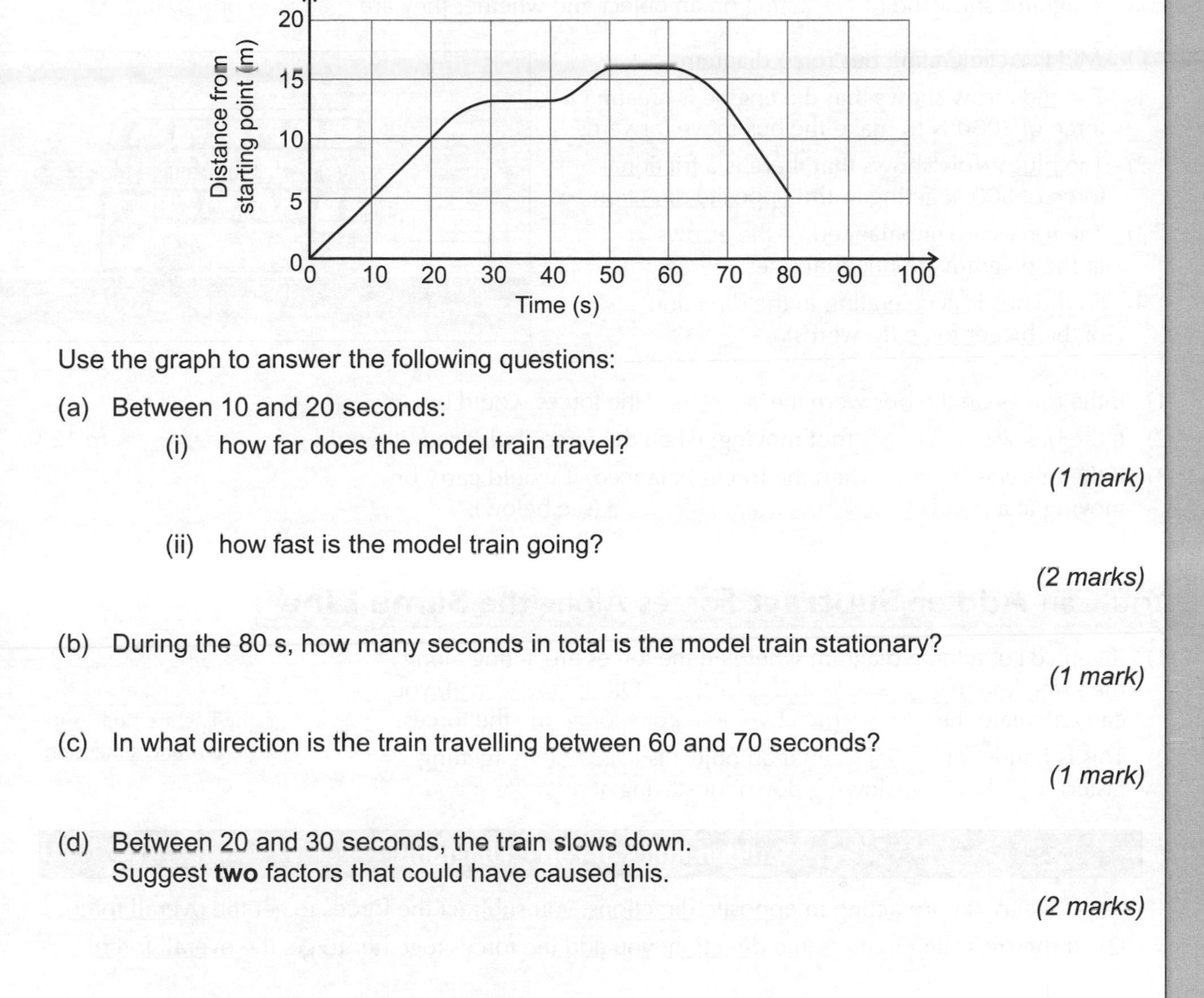

Use the graph to answer the following questions:

(a) Between 10 and 20 seconds:

(i) how far does the model train travel?

(1 mark)

(ii) how fast is the model train going?

(2 marks)

(b) During the 80 s, how many seconds in total is the model train stationary?

(1 mark)

(c) In what direction is the train travelling between 60 and 70 seconds?

(1 mark)

(d) Between 20 and 30 seconds, the train slows down. Suggest **two** factors that could have caused this.

(2 marks)

3 Three aircraft are taking part in a display.

(a) In one part of the display, two of the aircraft fly towards each other as shown:

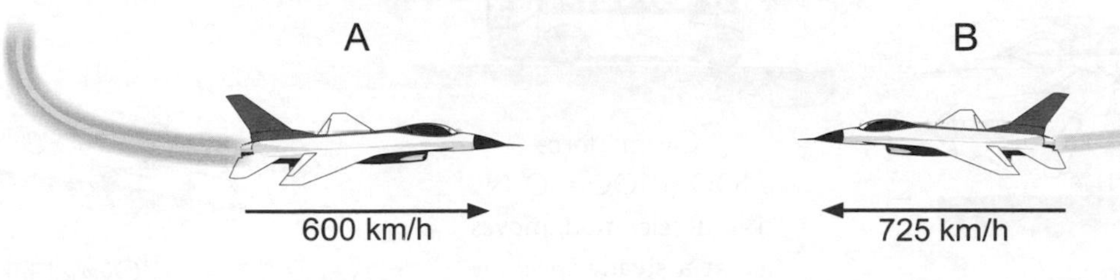

Calculate the speed of aircraft B relative to aircraft A.

(1 mark)

(b) The third aircraft is travelling in the same direction as aircraft A. How would you calculate its speed relative to aircraft A?

(1 mark)

Force Diagrams

Force diagrams. They're diagrams that show forces. Bet you weren't expecting that...

Show the Forces Acting on an Object Using a Force Diagram

Force diagrams show the forces acting on an object and whether they are balanced or unbalanced.

EXAMPLE: Accelerating bus force diagram

1) The red arrow shows that the engine is creating a force of 2000 N to make the bus move forwards.
2) The blue arrow shows that there is a frictional force of 500 N acting in the opposite direction.
3) The forces are unbalanced — the arrows in the diagram are unequal sizes.
4) So the bus is accelerating in the direction of the bigger force (forwards).

1) If the forces on the bus were the same size, the forces would be BALANCED.
2) If the bus was stationary (not moving) when the forces balanced, it would remain stationary (p.123).
3) If the bus was moving when the forces balanced, it would carry on moving at a steady speed in the same direction (see below).

You Can Add or Subtract Forces Along the Same Line

1) If you've got a force diagram where all the forces are acting along the same line (e.g. forwards and backwards OR up and down), you can calculate the overall force by adding or subtracting the forces.
2) This is handy for working out if an object is accelerating (getting faster), decelerating (slowing down) or staying at a steady speed:

Forces acting along the same line are said to be acting in one dimension.

Golden Rules of Force Diagrams

1) If the forces are acting in opposite directions, you subtract the forces to get the overall force.
2) If they're acting in the same direction, you add the forces together to get the overall force.

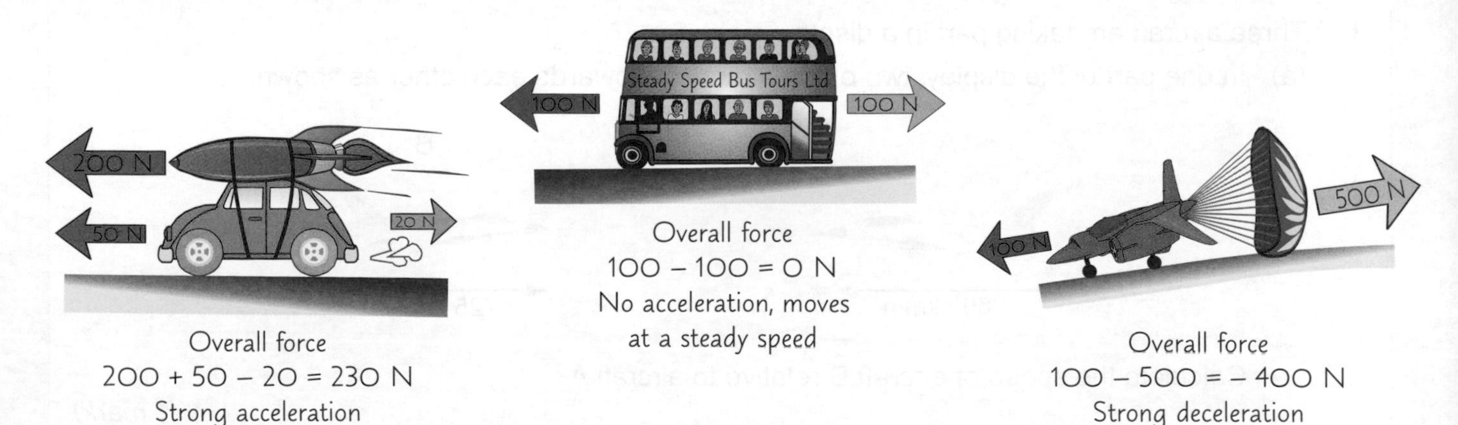

Overall force
200 + 50 – 20 = 230 N
Strong acceleration

Overall force
100 – 100 = 0 N
No acceleration, moves at a steady speed

Overall force
100 – 500 = – 400 N
Strong deceleration

Use the force — look at a diagram, find out if it's accelerating...

Remember: if the arrows on a force diagram are equal and opposite, the object isn't moving or is moving at a steady speed. If they're different sizes, the object will accelerate or decelerate.

Moments

Don't wait a lifetime to learn moments like this — memorise what's on this page now.

Forces Cause Objects to Turn Around Pivots

A pivot is the point around which rotation happens — like the middle of a seesaw.

A Moment is the Turning Effect of a Force

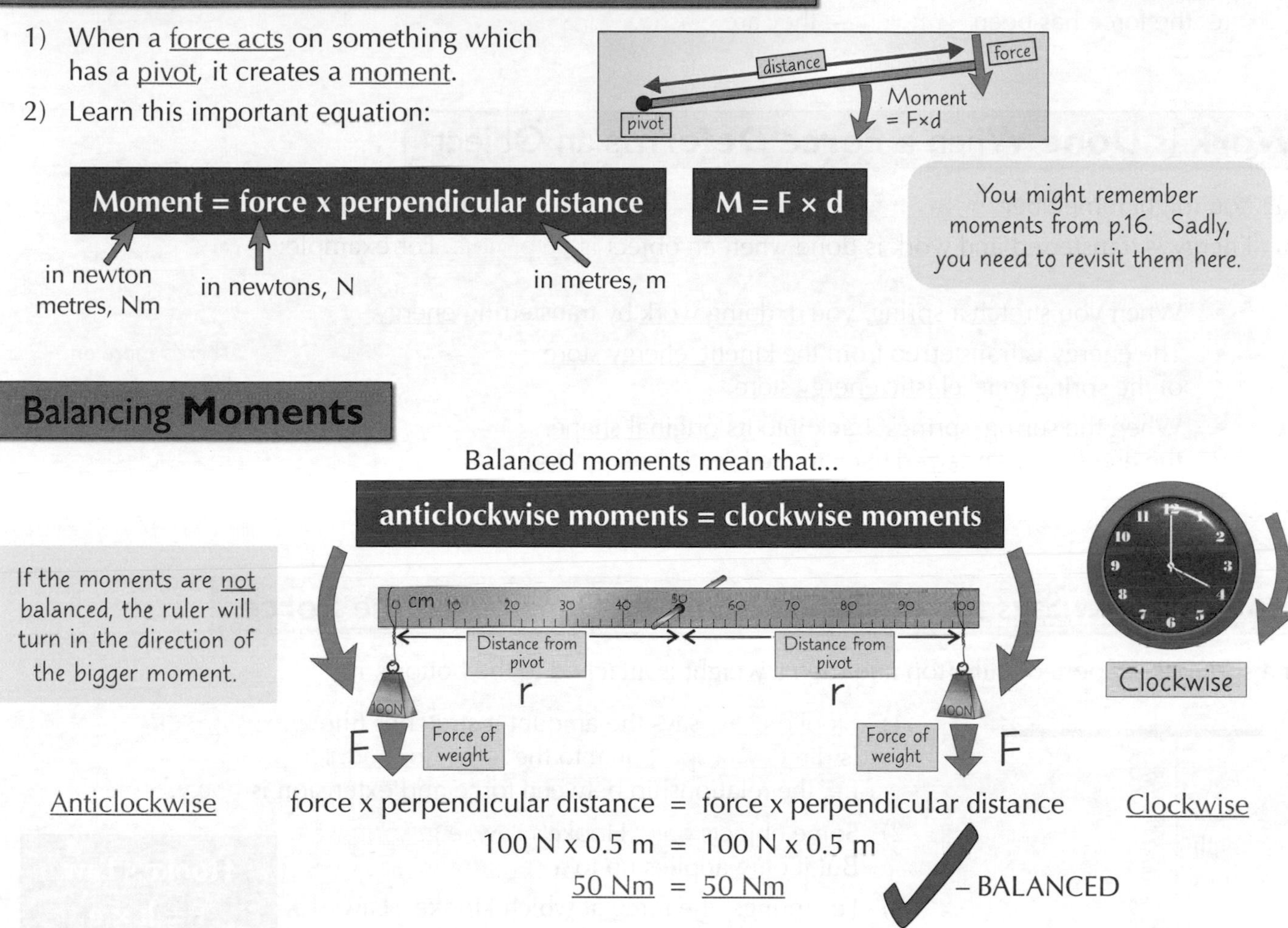

1) When a force acts on something which has a pivot, it creates a moment.
2) Learn this important equation:

Moment = force x perpendicular distance **M = F × d**

You might remember moments from p.16. Sadly, you need to revisit them here.

Balancing Moments

Balanced moments mean that...

anticlockwise moments = clockwise moments

If the moments are not balanced, the ruler will turn in the direction of the bigger moment.

Anticlockwise	force x perpendicular distance	=	force x perpendicular distance	Clockwise
	100 N x 0.5 m	=	100 N x 0.5 m	
	50 Nm	=	50 Nm	– BALANCED

Is it Balanced?

Which rulers are balanced? If you think the ruler is balanced write it on a bit of paper. If you reckon it's unbalanced, then write unbalanced, but say which side of the ruler will dip down. Words to use: balanced, unbalanced, left side down or right side down. Answers on page 196.

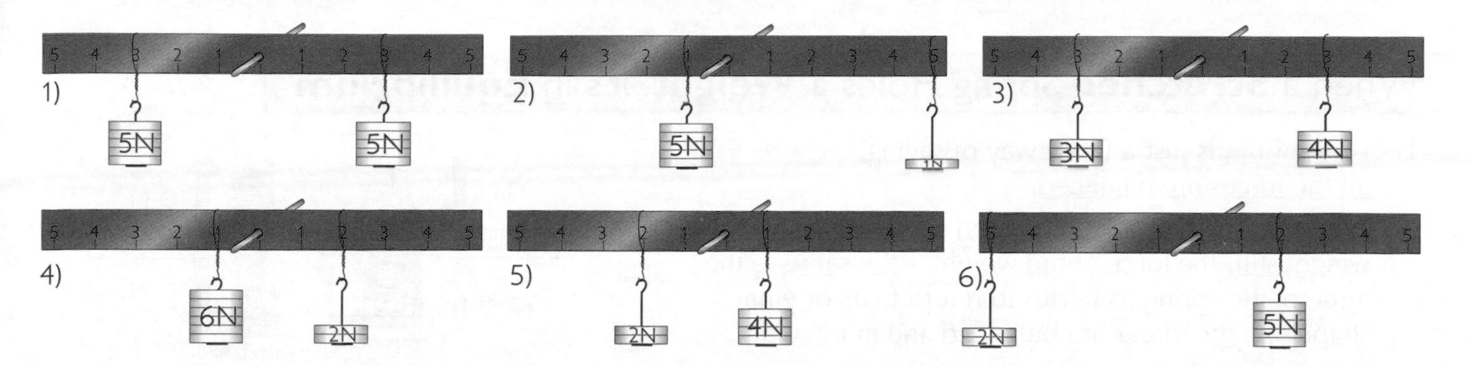

Forces and Elasticity

It's not just about turning, pushing and pulling — forces are also able to stretch or squash things.

You Can **Deform** Objects by **Stretching** or **Squashing**

1) You can use forces to stretch or compress (squash) objects, e.g. when you stomp on an empty fizzy pop can.
2) The force you apply causes the object to deform (change its shape).
3) Springs are special because they usually spring back into their original shape after the force has been removed — they are elastic.

Work is **Done** When a **Force Deforms** an Object

1) You might remember energy transfer from page 102 (if not, take a look). Work done is the same thing.
2) Energy is transferred and work is done when an object is deformed. For example:

- When you stretch a spring, you're doing work by transferring energy.
- The energy is transferred from the kinetic energy store of the spring to its elastic energy store.
- When the spring 'springs' back into its original shape, the energy is transferred back to the kinetic energy store.

There's more on different stores of energy on page 102.

Hooke's Law Says **Extension** of a **Spring** Depends on the **Force**

If a spring is supported at the top and then a weight is attached to the bottom, it stretches.

Natural length l

Extension, e

Force, F

1) Hooke's Law says the amount it stretches (the extension, e), is directly proportional to the force applied, F. I.e. the relationship between force and extension is linear.
2) Some objects obey Hooke's Law, e.g. springs. But it only applies up to a certain force.
3) For springs, the force at which Hooke's Law stops working is much higher than for most materials. Springs are unusual.

k is the spring constant. Its value depends on the material that you're stretching and it's measured in newtons per metre (N/m).

When a **Stretched** Spring Holds a **Weight**, it's in **Equilibrium**

1) Equilibrium is just a fancy way of saying all the forces are balanced.
2) When a stretched or compressed spring holds a weight still, the force of the weight is the same as the force of the spring as it tries to return to its original shape. So the forces are balanced and in equilibrium.

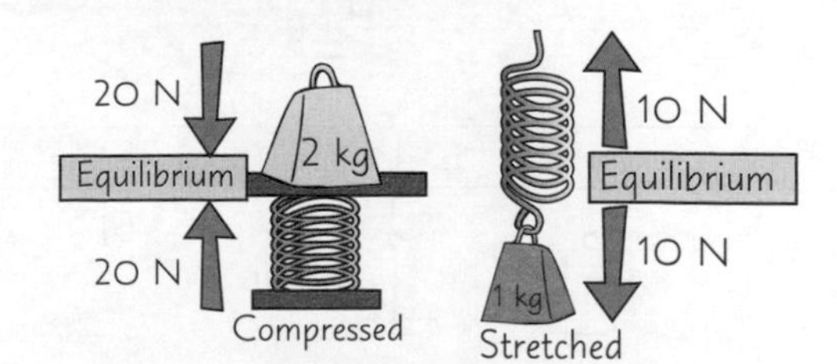

Pressure

Don't let pressure get you down — here's a lovely couple of pages that explains it all. Enjoy.

Pressure is How Much **Force** is Put on a Certain **Area**

Pressure, force and area are all kind of tied up with each other — as the formula shows. The formula can also be put in a triangle, which is nice.

A given force acting over a big area means a small pressure (and vice versa).

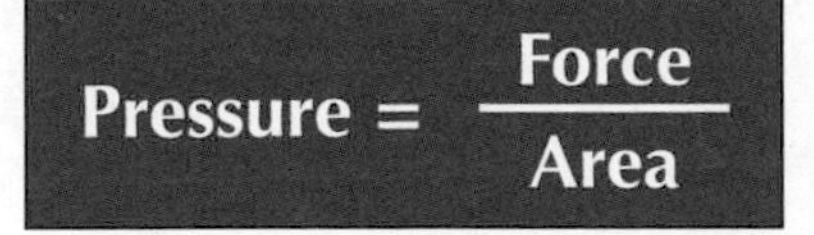

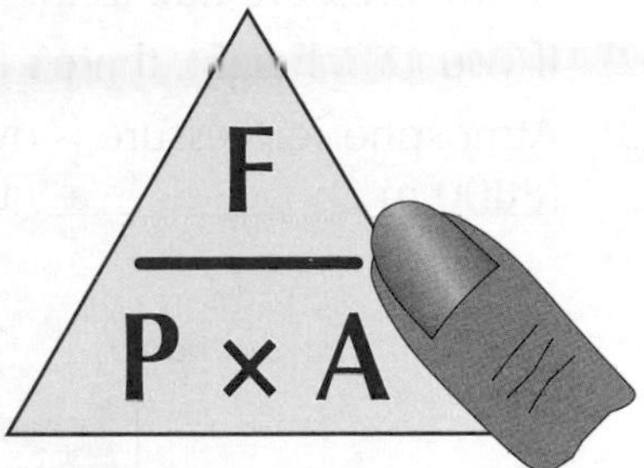

Pressure is Measured in **N/m²** or **Pascals** (Pa)

1 newton/metre² = 1 pascal
1 N/m² = 1 Pa

If a force of 1 newton is spread over an area of 1 m² (like this) then it exerts a pressure of 1 pascal. Simple as that.

1N

Force acts normal (at 90°) to area.

1m²

1N

1m

1m

Pressure = 1 Pa

Example: A wooden box weighs 15 N and its base has an area of 0.8 m². Calculate the pressure exerted by the box on the floor.

Answer: Pressure = force ÷ area.
15 ÷ 0.8 = 18.75 N/m² or 18.75 Pa.

15 N

MATHS TIP

Don't feel too pressured about this stuff

Rearranging equations and swapping between units can be tricky, but it gets easier with practice. Formula triangles (p.120) are great for when you have to do some rearranging.

Pressure

Atmospheric Pressure is All Around Us All the Time

The weight of the atmosphere is constantly pushing against us — but we're so used to it we can't feel it.

1) The lower you are, the more atmosphere there is above you — so the pressure due to the weight of the atmosphere increases.
2) If you gain height, there's less atmosphere above you, so the atmospheric pressure decreases.
3) Atmospheric pressure is over 100 000 Pa at sea level. But at the top of Mt Everest (8800 m above sea level) the atmospheric pressure is only around 33 000 Pa.

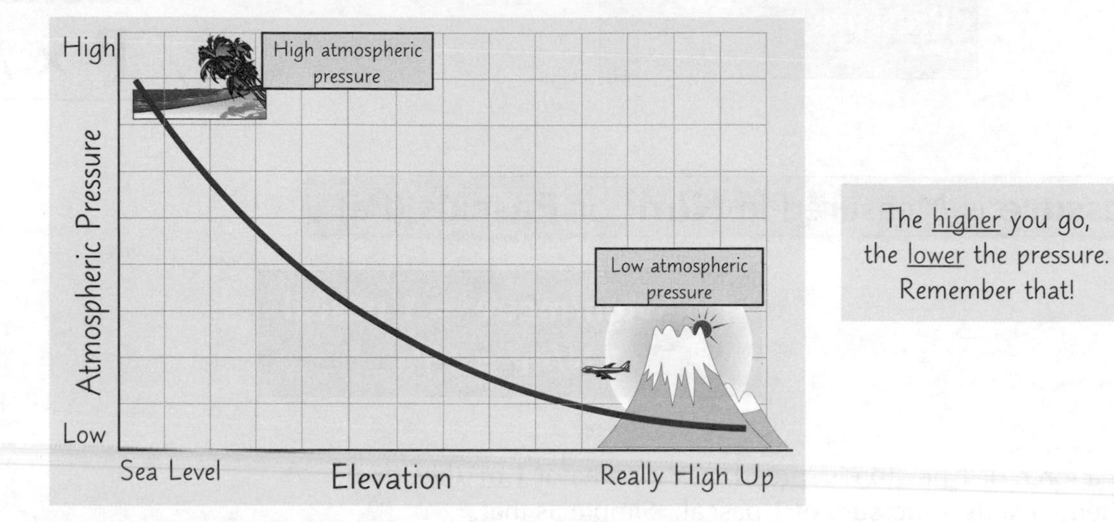

The higher you go, the lower the pressure. Remember that!

The Pressure in Liquids Increases with Depth

For liquids like water, the pressure increases with depth due to the weight of water above.

Water Pressure Causes Upthrust and Makes Things Float

1) If you place an object in water, it experiences water pressure from all directions.
2) Because water pressure increases with depth, the force pushing upwards at the bottom of the object is greater than the force pushing down at the top of the object.
3) This causes an overall upwards force, called upthrust.
4) If the upthrust is equal to the object's weight, then the object will float — like this boat:
5) If the upthrust is less than the object's weight, it will sink.

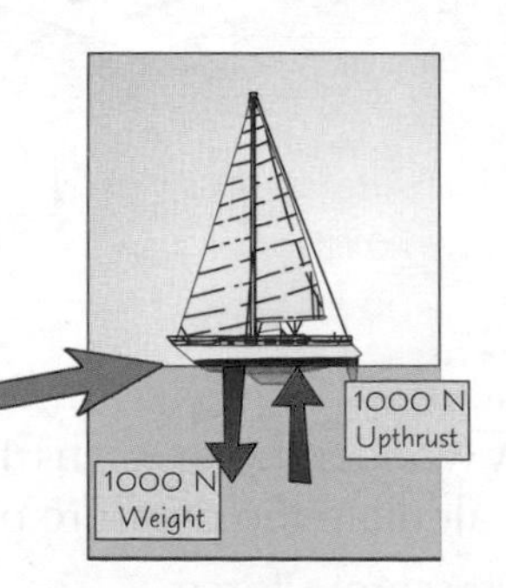

Pressure — it pushes down on you all the time

Remember: air and water have a mass. If you increase the amount of this mass above your head, you're increasing the force acting over a given area and you'll get an increase in pressure. That's why pressure in liquids increases with depth and why air pressure decreases with height. Re-read this page to make sure you really understand it before moving on.

Warm-Up and Practice Questions

There are quite a lot of calculations to get to grips with here.
Go back over any ones you get wrong until you're confident about how to do them.

Warm-Up Questions

1) Draw a force diagram for a car moving at a steady speed with a driving force of 2000 N. Ignore the vertical forces on the car.
2) What is the overall force on a motorbike if the engine provides a driving force of 1000 N and it feels resistive forces of 400 N?
3) Fill in the missing words in the following sentence. If the moments on an object are balanced, you know that: __________ moments = __________ moments.
4) What does it mean if a stretched spring is in equilibrium?
5) A girl weighing 500 N balances on one ice-skate. The area of the skate in contact with the ice is 0.002 m^2. How much pressure does she exert on the ice? Give the units.
6) In terms of forces, why does a dense object sink when placed in water?

Practice Questions

1 Arnold attaches a mass to a fixed spring and allows it to fall.

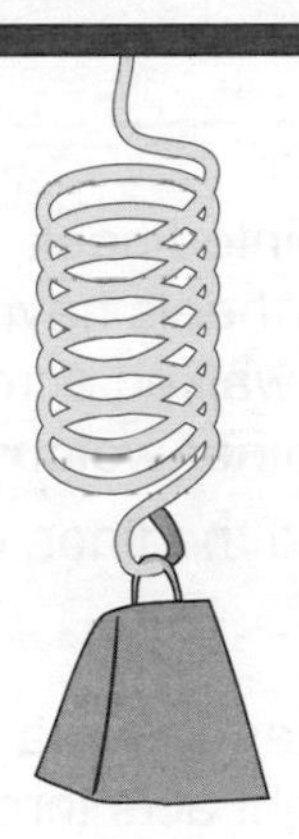

(a) As the mass falls, energy is transferred to its kinetic energy store. As the spring stretches, the mass slows down and the energy in its kinetic energy store is reduced. What type of energy store is the energy transferred to?

(1 mark)

(b) The spring extends by a total of 0.2 metres. Use Hooke's Law to calculate the weight of the mass. The spring constant k = 100 N/m.

(1 mark)

(c) Eventually the mass comes to rest and is held in equilibrium by the spring. What is the size of the force being exerted on the mass by the spring?

(1 mark)

Practice Questions

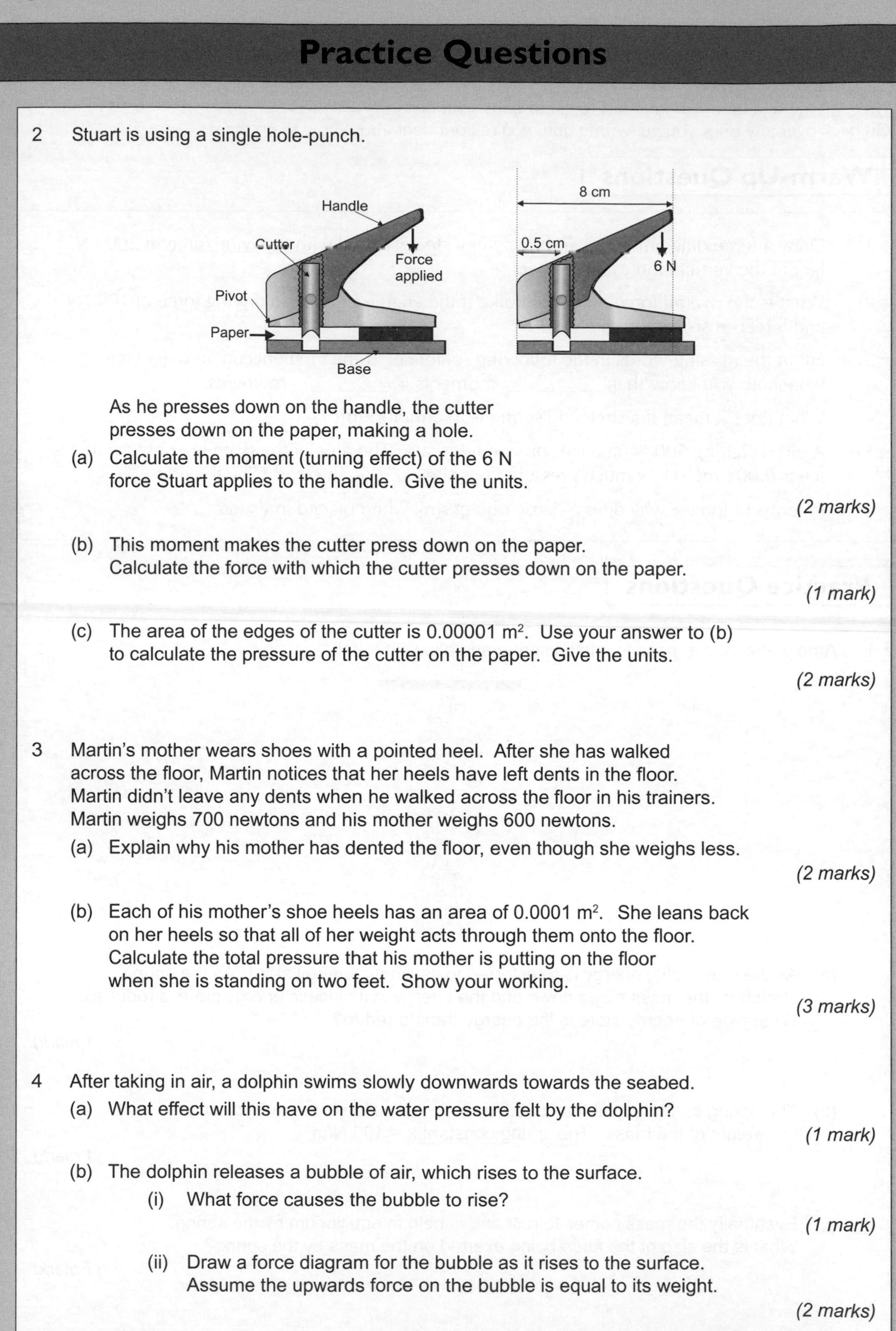

2 Stuart is using a single hole-punch.

As he presses down on the handle, the cutter presses down on the paper, making a hole.

(a) Calculate the moment (turning effect) of the 6 N force Stuart applies to the handle. Give the units.

(2 marks)

(b) This moment makes the cutter press down on the paper. Calculate the force with which the cutter presses down on the paper.

(1 mark)

(c) The area of the edges of the cutter is 0.00001 m^2. Use your answer to (b) to calculate the pressure of the cutter on the paper. Give the units.

(2 marks)

3 Martin's mother wears shoes with a pointed heel. After she has walked across the floor, Martin notices that her heels have left dents in the floor. Martin didn't leave any dents when he walked across the floor in his trainers. Martin weighs 700 newtons and his mother weighs 600 newtons.

(a) Explain why his mother has dented the floor, even though she weighs less.

(2 marks)

(b) Each of his mother's shoe heels has an area of 0.0001 m^2. She leans back on her heels so that all of her weight acts through them onto the floor. Calculate the total pressure that his mother is putting on the floor when she is standing on two feet. Show your working.

(3 marks)

4 After taking in air, a dolphin swims slowly downwards towards the seabed.

(a) What effect will this have on the water pressure felt by the dolphin?

(1 mark)

(b) The dolphin releases a bubble of air, which rises to the surface.

(i) What force causes the bubble to rise?

(1 mark)

(ii) Draw a force diagram for the bubble as it rises to the surface. Assume the upwards force on the bubble is equal to its weight.

(2 marks)

Revision Summary for Section Nine

Section Nine is all about forces and motion. It's all pretty straightforward stuff really and the questions below will test whether you've learnt the basic facts.

If you're having trouble learning the stuff, try taking just one page on its own. Start by learning part of it, then covering it up and scribbling it down again. Then learn a bit more and jot that down. Soon enough you'll have learnt the whole section and be ready to face any question your teachers throw at you.

1) What exactly is speed? Write down the formula triangle for speed.
2) How does SIDOT help you remember what speed is?
3)* A piece of paper is flicked across the lab by a student. It travels 5 m in 2 seconds. Calculate the speed of the paper.
4)* On sports day you run 100 m in 20 seconds. Can you run faster than the flicked piece of paper?
5)* When a car is going at 40 mph, how far will it travel in 15 minutes?
6) What does the gradient show on a distance-time graph?
7) What does a straight, flat line mean on a distance-time graph?
8) How would you calculate the relative speed of two trains travelling in the same direction?
9) Can forces be seen? How do we know they're there?
10) What are the units of force? What would you use to measure force?
11) What are the five different things that forces can make objects do?
12) What do balanced forces produce? What do unbalanced forces do?
13) What is friction? When does it occur?
14) Give three good points of friction. Give two bad points of friction.
15) What is air resistance? And water resistance?
16) When a sheep first jumps out of a plane what happens to its speed?
17) As the sheep moves faster, what happens to the air resistance?
18) What happens to air resistance when the sheep's parachute opens?
19) Does the speed then change? When does the sheep's speed become steady?
20) What might happen if the ground didn't provide an upward force to equal the sheep's weight?
21) Draw a force diagram of a kettle resting on a table. The force due to gravity acting on the kettle is 10 N.
22) If the forces acting on a moving bus are balanced, what will happen to its speed?
23) What is a moment? Give the formula for a moment.
24) What does "balanced moments" mean?
25)* A weight of 100 N is put 1 m away from the middle of a horizontal seesaw. What distance from the middle should a weight of 50 N be applied to balance the seesaw?
26) Give two ways you can deform objects.
27) What does Hooke's Law say? Write down the formula.
28) What is pressure? Give the formula for calculating pressure.
29)* A force of 200 N acts on an area of 2 m^2. Calculate the pressure.
30) Is atmospheric pressure higher at the seaside or up a mountain? Why?
31) When does an object placed in water float?

*Answers on page 197

Water Waves

Water waves transfer energy from one place to another. You can see them at the beach.

Water Waves are **Transverse**

1) Waves travelling across the ocean are good examples of transverse waves.
2) A transverse wave has undulations (up and down movements) that are at right angles to the direction the wave is travelling in.

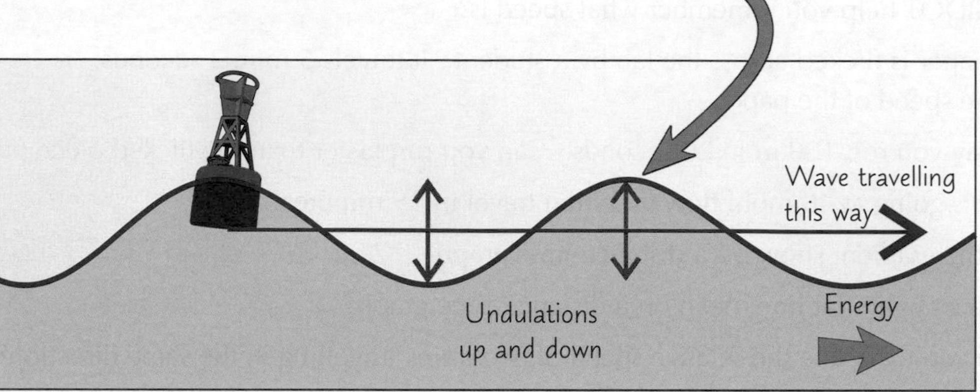

3) Waves transfer energy from one place to another. So the undulations are also at right angles to the direction of energy transfer.
4) Lots of other important waves are transverse too, like light (see page 138).

Waves Can be **Reflected**

1) If a water wave hits a surface, it will be reflected.
2) This causes the direction of the wave to change.
3) All waves can be reflected. There's more on reflection on page 139.

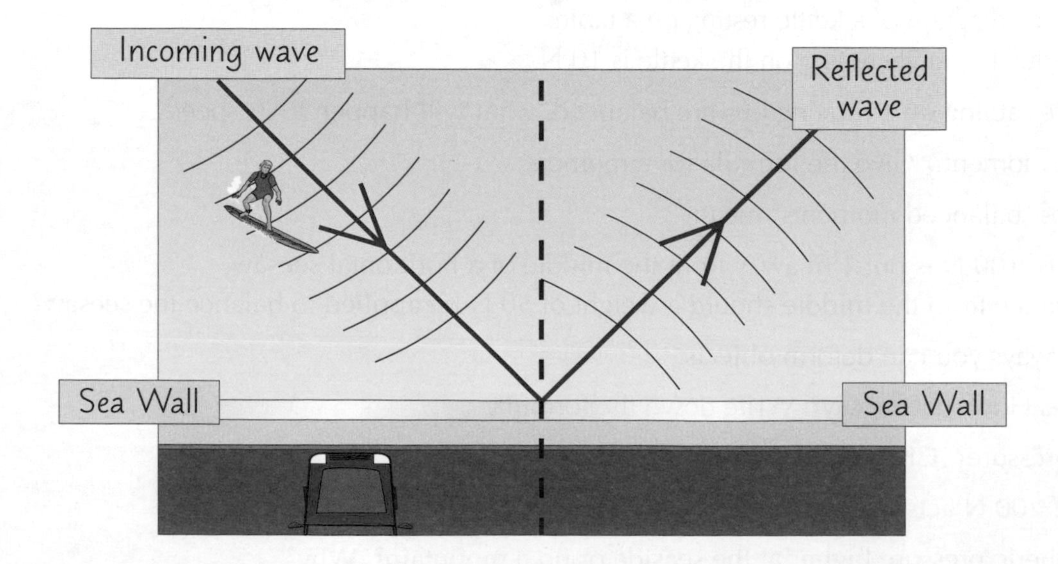

Water Waves

Transverse Waves Have Crests, Troughs and Displacement

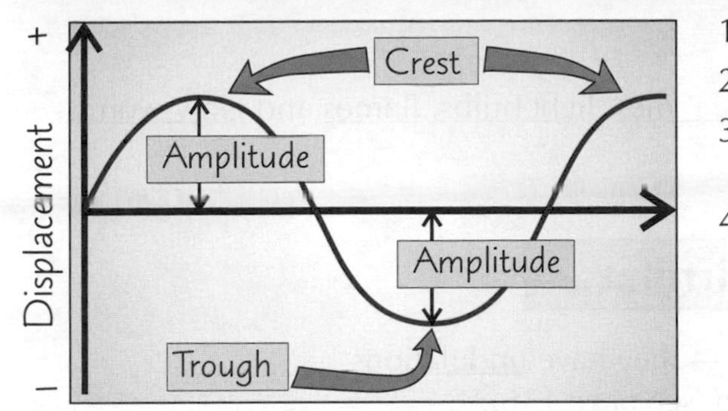

1) The crest is the highest part of the wave.
2) The trough is the lowest part of the wave.
3) The displacement is how far a point on the wave is from the middle line.
4) The amplitude is the maximum displacement — the distance from the middle of the wave to a crest or trough.

Superposition Happens When Two Waves Meet

1) If two water waves meet, their displacements combine briefly. This is superposition. The waves then carry on as they were.
2) If two identical crests meet, the height of each wave is added together. So the crest height doubles.

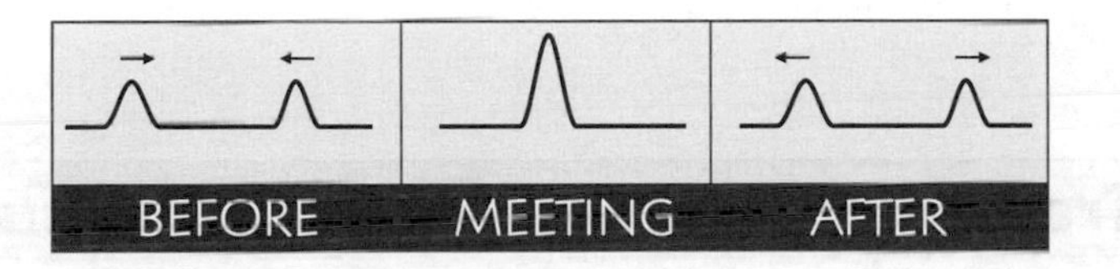

3) If two identical troughs meet, the depth of each wave is added together. So the trough depth doubles.

4) If one wave is at a crest and the other is at a trough, you subtract the trough depth from the crest height. So the crest or trough will be smaller and may even cancel out, leaving a flat water surface.

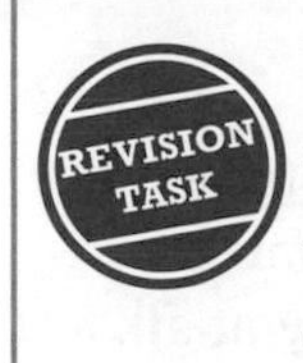

Crests, troughs, amplitude, displacement — learn the labels

Waves are a bit tricky. So grab a piece of paper and write down all the technical terms on these pages and what they mean. Then draw and explain the diagrams. Finally, do all that again with the book closed until you've memorised both pages — even the difficult bits...

Light Waves

You wouldn't know from looking, but light is actually a wave. Here's a page all about it...

Light is a **Wave** that Transfers **Energy**

1) Light is produced by luminous objects such as the Sun, candles, light bulbs, flames and glow worms.
2) Light is a wave, which always travels in a straight line.

Light Waves and **Water Waves** Are **Similar**...

1) Like waves in water, light waves are transverse waves — they have undulations at right angles to the direction the wave is travelling in (see page 136).
2) And like waves in water, light waves transfer energy from one place to another.
3) Light waves can be reflected too — this is how mirrors work (see next page for more).

...But **Light** Waves **Don't Need Particles** to Travel

1) Water waves travel (and transfer energy) by moving particles.
2) Light waves don't need particles to travel. This is a good thing — light from the Sun has to travel through space (where there aren't many particles, see below) to reach Earth.
3) Light waves are slowed down by particles.

Light Waves **Always Travel** at the **Same Speed** in a **Vacuum**

1) Light travels faster when there are fewer particles to get in the way.
2) Light always travels fastest in a vacuum. A vacuum is where there is nothing at all — no air, no particles, nothing. Space is mostly a vacuum.
3) The speed of light in a vacuum is always 3×10^8 m/s (that's three hundred million metres per second). It's a constant.
4) This means light from the Sun gets to Earth in only 8.3 minutes — even though it's 150,000,000 km away.
5) Nothing travels faster than light in a vacuum.
6) Make sure you learn this:

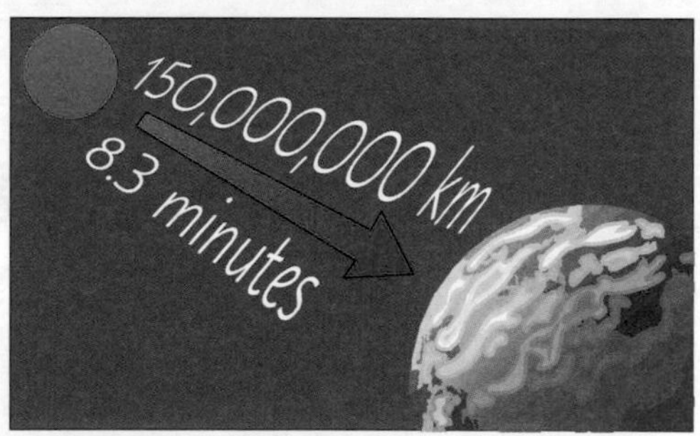

Speed of light waves in a vacuum = 3×10^8 m/s

7) Although light is slower when it has to travel through matter (like air or water), it's still so fast that its movement appears instant to the human eye.

Light waves are similar to water waves — but not the same

Light is just like all those waves you see at the beach. Except that it doesn't need a load of water to get from A to B — anything like water puts a load of particles in the way and slows the light waves down. Nope, light only hits top gear when it's in a vacuum with absolutely nothing in the way at all.

Reflection

Reflection is what happens when a wave hits a surface.

Mirrors Have **Shiny Surfaces** Which **Reflect Light**

1) A light wave is also known as a light ray.
   Light rays reflect off mirrors and most other things.
2) Mirrors have a very smooth shiny surface, which reflects all the light off at the same angle, giving a clear reflection. This is specular reflection.
3) Rough surfaces look dull, because the light is reflected back (scattered) in lots of different directions. This is diffuse reflection (or diffuse scattering).

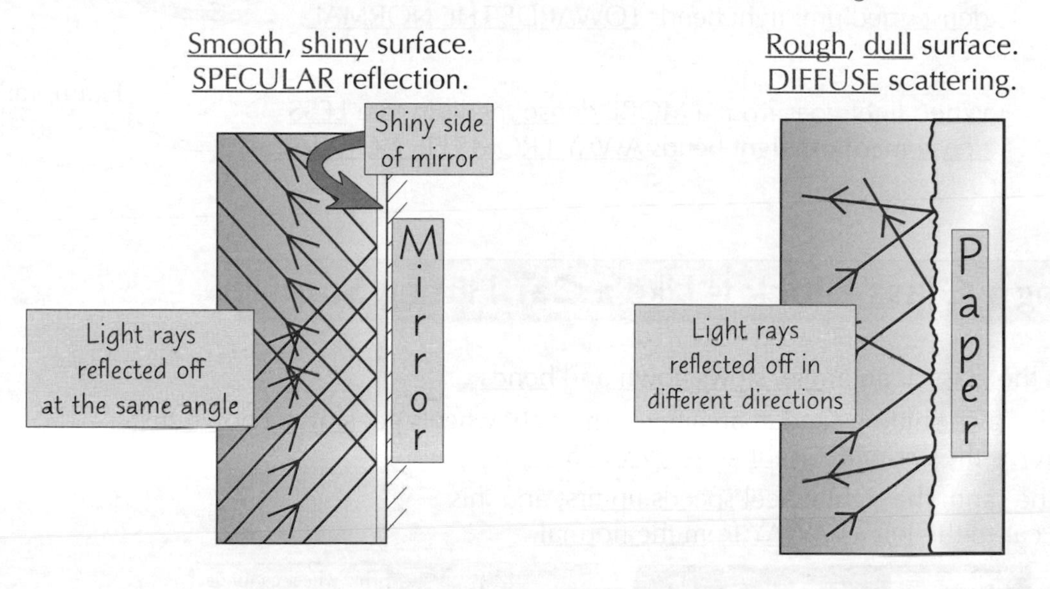

Learn the **Law** of **Reflection:**

Angle of incidence = angle of reflection
Angle i = angle r

1) The angle of incidence and the angle of reflection are always measured between the light ray and the normal.
2) The normal is a line at a right angle (90°) to the surface.

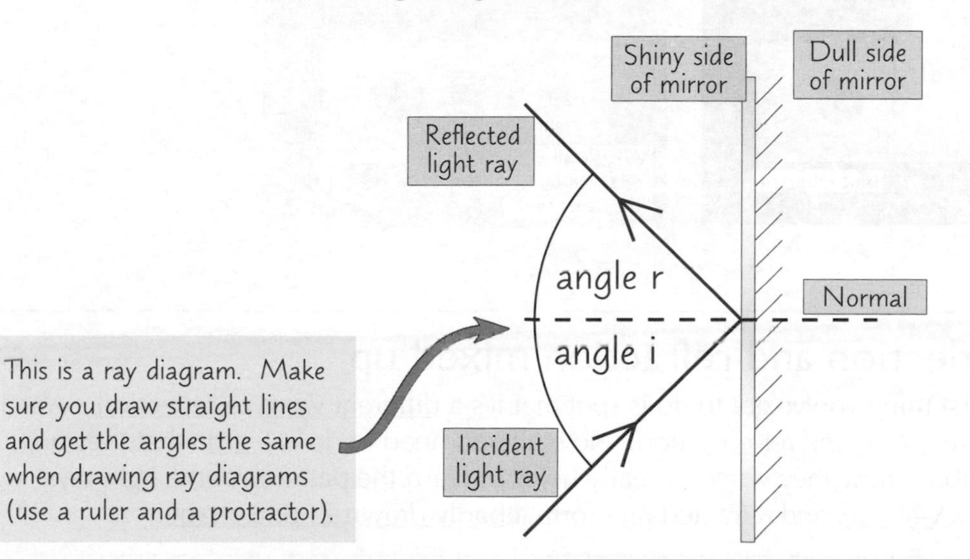

Refraction

Refraction happens when light rays move to a more or less dense substance.

Refraction is When Light Bends as it Crosses a Boundary

1) Light will travel through transparent (see-through) material, but it won't go through anything opaque (not see-through).
2) Any substance that light (or another wave, e.g. sound) travels through is called a medium.
3) When light travels from one transparent medium to another, it bends or refracts.

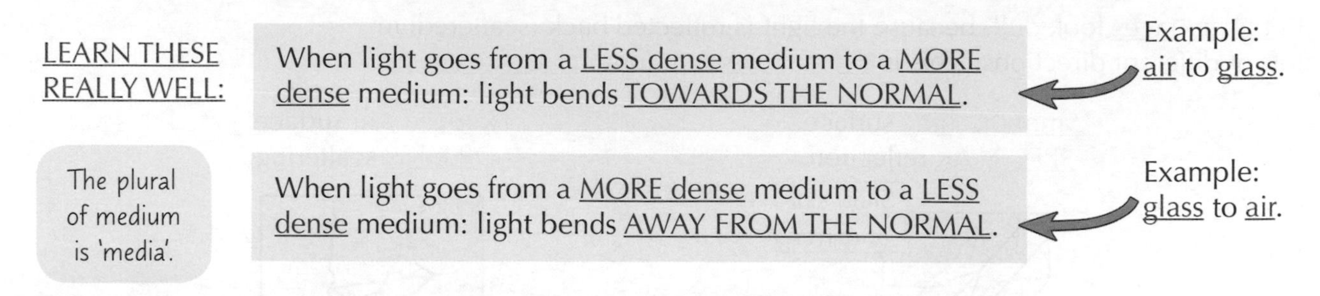

LEARN THESE REALLY WELL:

When light goes from a LESS dense medium to a MORE dense medium: light bends TOWARDS THE NORMAL. ← Example: air to glass.

When light goes from a MORE dense medium to a LESS dense medium: light bends AWAY FROM THE NORMAL. ← Example: glass to air.

The plural of medium is 'media'.

Light Hitting a Glass Block is Like a Car Hitting Sand

1) Light hits the glass at an angle, slows down and bends.
2) It's a bit like a car hitting sand at an angle. The right wheels get slowed down first and this turns the car to the right — TOWARDS the normal.
3) Leaving the sand, the right wheel speeds up first and this turns the car to the left — AWAY from the normal.

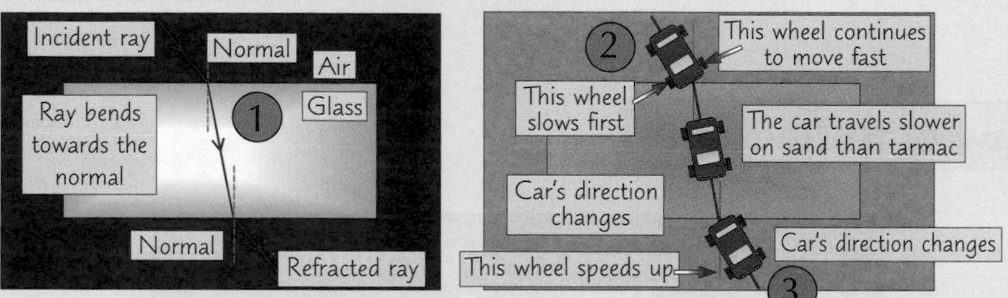

4) If both wheels hit the sand together they slow down together, so the car goes straight through, WITHOUT TURNING.
5) Light does exactly the same when it hits the glass block straight on.

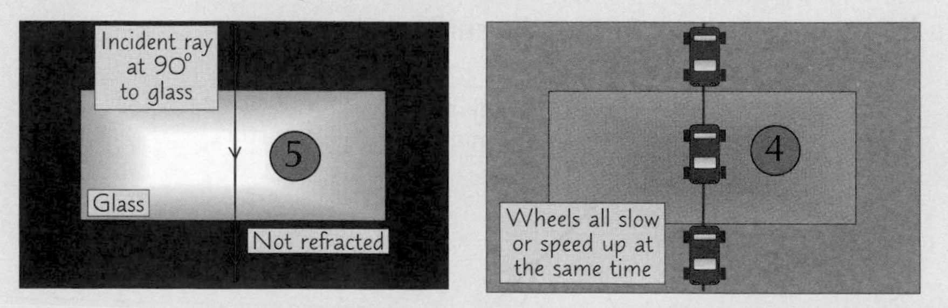

Don't get reflection and refraction mixed up

Refraction — the first thing you've got to do is spot that it's a different word to reflection. Watch, I'll do it again: ref-l-e-c-tion and ref-r-a-c-tion. Now all you need to do is learn what they are, and all the details about how they work. It really helps to learn the patterns of the light rays — reflected rays form a V-shape, and refracted rays form a badly drawn Z- or S-shape.

How We See

We See Things Because Light Reflects into our Eyes

1) When luminous objects produce light (see page 138), it reflects off non-luminous objects, e.g. you, me, books, sheep, etc.
2) Some of the reflected light then goes into our eyes and that, my friend, is how we see.

The Pinhole Camera is a Simple Camera

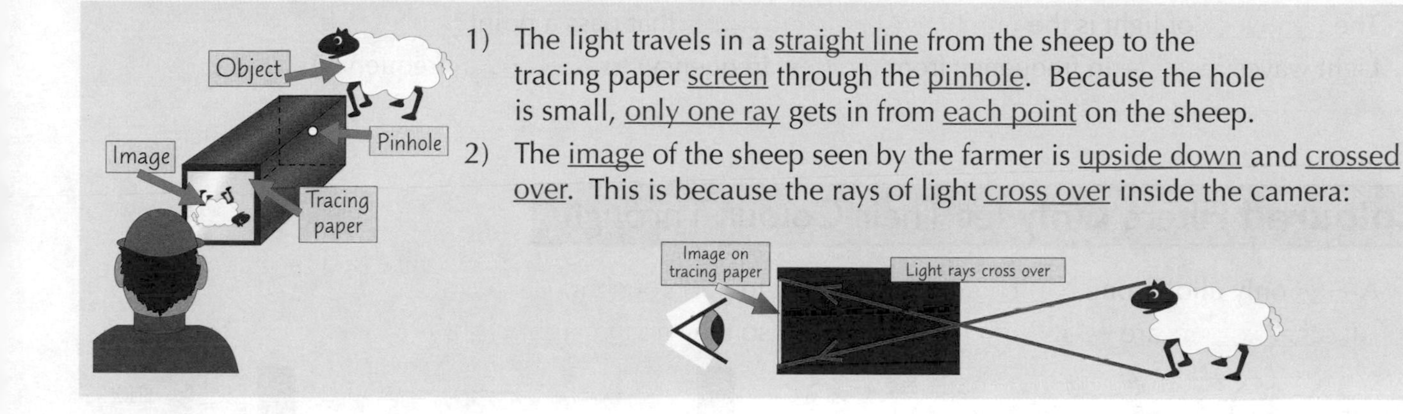

1) The light travels in a straight line from the sheep to the tracing paper screen through the pinhole. Because the hole is small, only one ray gets in from each point on the sheep.
2) The image of the sheep seen by the farmer is upside down and crossed over. This is because the rays of light cross over inside the camera:

Lenses Can be Used to Focus Light

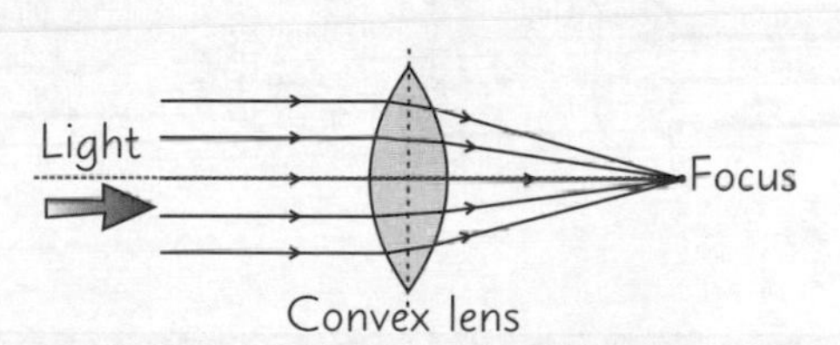

1) A lens refracts (bends) light.
2) A convex lens bulges outwards. It causes rays of light to converge (move together) to a focus.

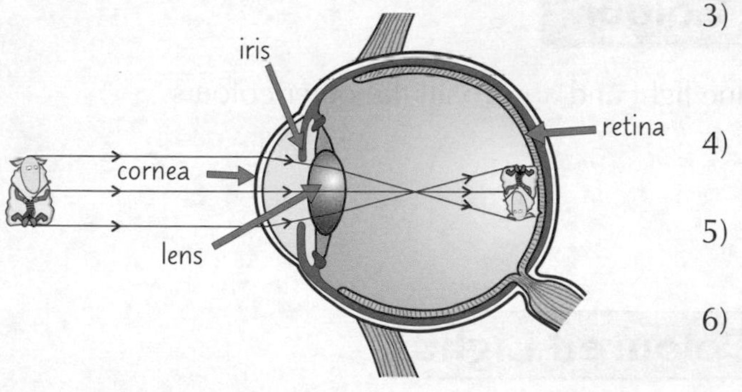

3) In the human eye, the cornea is a transparent 'window' with a convex shape. The cornea does most of the eye's focusing.
4) The convex lens behind the cornea changes shape to focus light from objects at varying distances.
5) The iris is the coloured part of the eye. It controls the amount of light entering the eye.
6) Images are formed on the retina. Cells in the retina are photo-sensitive (sensitive to light).

Energy is Transferred from a Light Source to an Absorber by Light

1) Energy is carried by light waves.
2) Anything that absorbs this energy is called an absorber, e.g. a retina cell in the eye, the film in a film camera or the digital image sensor in a digital camera.
3) The energy is transferred to the absorber when it hits the absorber.
4) When light waves hit a retina cell it causes chemical and electrical changes in special cells that send signals to the brain.
5) In a digital camera, light causes the sensor to generate an electrical charge. The changes in charge are read by a computer and turned into an image.

Colour

White Light is Not just a Single Colour

1) Bit of a shocker, I know — but white light is actually a mixture of colours.
2) This shows up bigstyle when white light hits a prism or a raindrop. It gets dispersed (i.e. split up) into a full rainbow of colours.
3) The proper name for this rainbow effect is a spectrum.
4) Learn the order that the colours come out in:
   Red Orange Yellow Green Blue Indigo Violet
   Remember it with this historical jollyism:
   Richard Of York Gave Battle In Velvet

5) The frequency of light is the number of complete waves that pass a point per second.
6) Light waves increase in frequency from red (low frequency) to violet (high frequency).

Coloured Filters Only let Their Colour Through

1) A filter only allows one particular colour of light to go through it.
2) All other colours are ABSORBED by the filter — so they don't get through.

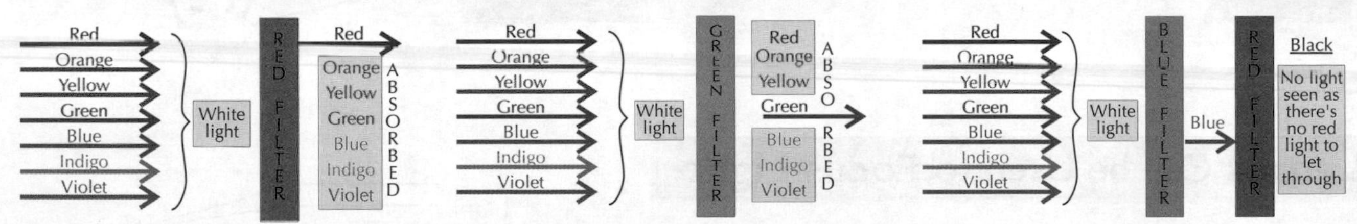

Coloured Objects Reflect Only That Colour

1) Blue jeans are blue because they diffusely reflect blue light and absorb all the other colours.
2) White objects REFLECT all colours.
3) Black objects ABSORB all colours.

Objects Seem to Change Colour in Coloured Light

IN WHITE LIGHT

White light

1) The boot looks red — it reflects red light and absorbs all other colours.
2) The lace looks green — it reflects green light and absorbs all other colours.

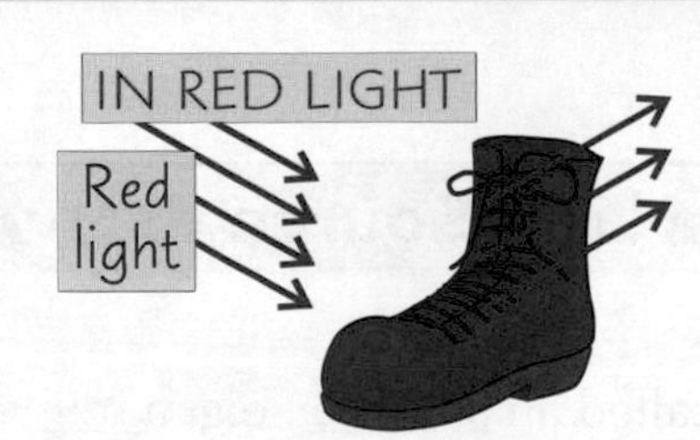

1) The boot looks red — it reflects the red light.
2) The lace looks black — it has no green light to reflect and it absorbs all the red light.

1) The boot looks black — it has no red light to reflect and it absorbs the green light.
2) The lace looks green — it reflects the green light.

Warm-Up and Practice Questions

You're now halfway through Section Ten. Time for a quick breather and a chance to check your understanding of the pages so far. Give these questions a go and make sure you haven't missed anything.

Warm-Up Questions

1) What type of wave are water waves?
2) True or false? The undulations of a transverse wave are in the same direction as the wave travels.
3) Explain what 'amplitude' means.
4) Explain why a convex lens can be used to focus light.
5) What colour of light has the lowest frequency?
6) Give an example of an object that can split up light into a spectrum of colours.

Practice Questions

1 Tyrone had his favourite green T-shirt and blue jeans on for the school disco. He also wore red boots with white laces.

(a) When he got to the disco there were red spotlights on the dance floor and green ones in the seating area. Copy and complete the table below to show what colour Tyrone's clothes seemed to be:

Tyrone's clothes	colour in red light	colour in green light
green T-shirt		
blue jeans		
red boots		
white laces		

(4 marks)

(b) At the end of the disco the DJ put the main white lights on, so that everybody could see and leave the school safely.

(i) What colour did Tyrone's jeans now appear to be?

(1 mark)

(ii) Why did they appear to be this colour?

(1 mark)

2 (a) Explain the role of each of these parts of the eye in helping us to see:

(i) cornea (ii) lens (iii) iris

(3 marks)

(b) (i) Describe the function of the retina in the eye.

(2 marks)

(ii) Which part of a digital camera plays the role of the retina in the human eye?

(1 mark)

Practice Questions

3 Bruno is investigating the properties of light in the lab.

(a) He shines a ray of light at a piece of paper.
Explain why the paper looks dull.

(2 marks)

(b) Bruno replaces the paper with a mirror. It gives a clear reflection.

(i) What name is given to this kind of reflection?

(1 mark)

(ii) He measures the angle of incidence of the ray of light on the mirror as 39°.
What is the angle of reflection?

(1 mark)

(c) Next, Bruno places a block of glass with parallel sides in the path of the ray of light.
He notices that the light bends when it enters the glass block, and bends again when it exits the far side of the glass block.

(i) The glass block is much more dense than air. In which direction does the light bend when it enters the glass block?

(1 mark)

(ii) Sketch a ray diagram to show the light entering and leaving the block.

(2 marks)

(d) (i) What type of wave are light waves?

(1 mark)

(ii) How fast does light travel in a vacuum?

(1 mark)

(iii) What happens to the speed of light when it travels from a vacuum into matter?

(1 mark)

4 (a) Describe what happens to the displacement of two water waves when:

(i) two crests with identical displacements meet.

(1 mark)

(ii) a trough with a depth of 2 cm and a crest with a height of 2 cm meet.

(1 mark)

(b) Give the name of the effect that causes the results you described in part (a).

(1 mark)

Sound

Like light, sound is a wave. It's a different type of wave to light though.

Longitudinal Waves Have Vibrations Along the Same Line

1) Longitudinal waves have vibrations that are parallel to the direction of the wave.
2) This means the vibrations are also parallel to the direction of energy transfer.
3) Examples of longitudinal waves include:
   - Sound waves.
   - A slinky spring when you push the end.

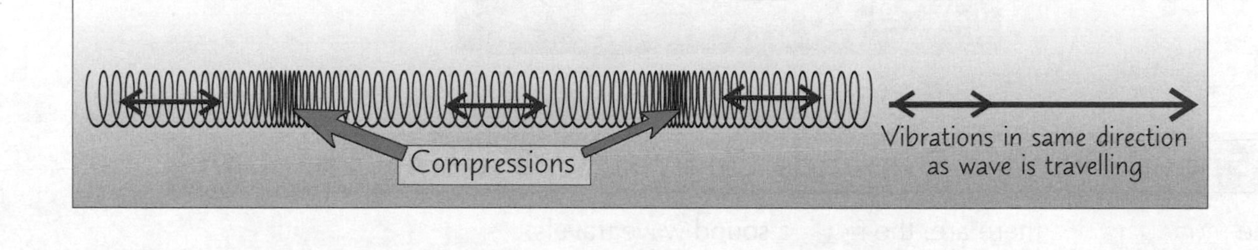

Sound Travels as a Longitudinal Pressure Wave

1) Sound waves are caused by vibrating objects.
2) Sound needs a medium (e.g. air or water) to travel through because something has to pass on the sound vibrations.

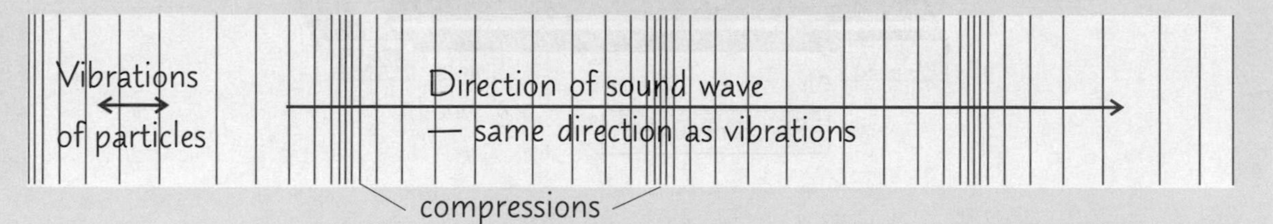

3) The vibrations are passed through the medium as a series of compressions (regions of squashed up particles).
4) Sound can't travel in space, because it's mostly a vacuum (there are no particles).

Learn the headings and you're halfway there

So make sure you do that first, then learn the diagrams and scribble them down from memory. Then do the same with the details. If you do that for five minutes, it might just help.

Sound

Sound Can be Reflected and Absorbed

1) Sound can be reflected and refracted just like light (see pages 139-140). An echo is sound being reflected from a surface.
2) Sound can also be absorbed. Soft things like carpets, curtains, etc. absorb sound easily.

The Speed of Sound Depends On What it's Passing Through

1) The more particles there are, the faster a sound wave travels.
2) Dense media have lots of particles in a small space. So the denser the medium, the faster sound travels through it (usually).
3) Sound generally travels faster in solids (like wood) than in liquids (like water) — and faster in liquids than in gases (like air).
4) Sound travels much slower than light.

Frequency is the Pitch of Sound

1) The frequency of sound is the number of complete waves that pass a point per second. A high frequency means more vibrations per second.
2) Frequency is a measure of how high-pitched (squeaky) the sound is. A high frequency means a high-pitched sound.
3) Frequency is measured in hertz (Hz) — the number of vibrations per second.

The speed of sound depends on what it's travelling through

Sound waves are caused by vibrations — if you've ever put a hand on a bass speaker (or turned up the volume and felt the floor vibrate) you'll have 'felt' a sound wave being made. Higher-pitched sounds are just the same, but the vibrations are more frequent.

Hearing

You might be wondering how vibrations in a sound wave turn into something you can hear. Read on...

Sound Waves Make Your Ear Drum Vibrate

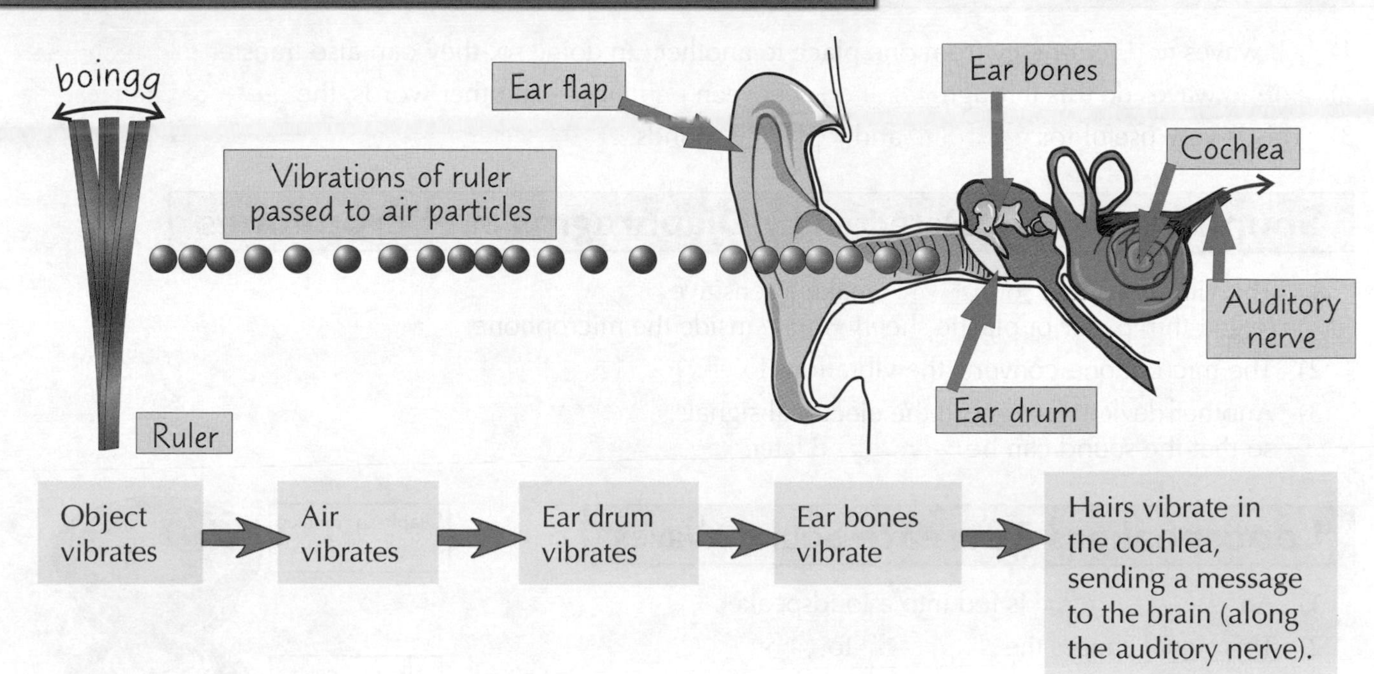

People and Animals Have Different Auditory Ranges

Your auditory range is the range of frequencies (vibrations per second) that you can hear.

1) The auditory range of humans varies a lot — but it's typically 20-20 000 Hz.
2) This means we can't hear low-pitched sounds with frequencies of less than 20 Hz or high-pitched sounds above 20 000 Hz.
3) Some animals like dogs, bats and dolphins can hear much higher frequencies than humans, as the chart shows.

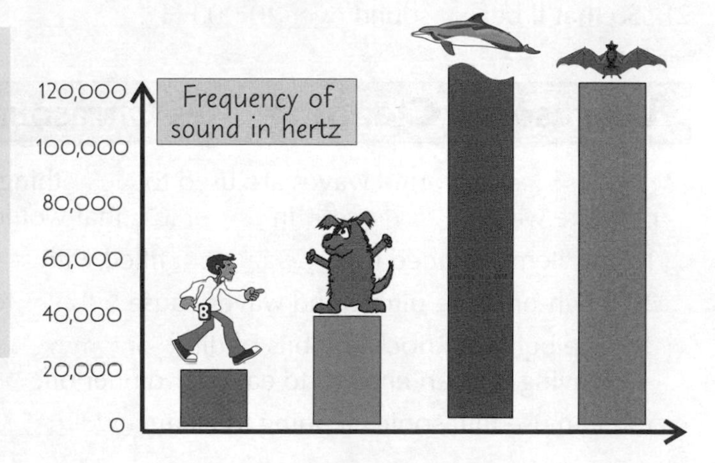

Look back at the previous page for more on frequency and pitch.

There's a fair bit of Biology on this page

The thing is, you still have to learn it. Remember, we hear things because the air carries the vibrations right into our ear. Learn the stuff about audible ranges for pitch too — it's pretty important.

Energy and Waves

You might remember how waves transfer energy (page 136). Here's a whole page on how useful that is.

Information Can be **Transferred** by **Pressure** Waves

1) All waves transfer energy from one place to another. In doing so, they can also transfer information.
2) Sound waves do this through vibrations between particles — in other words, the pressure changes.
3) This is very useful for recording and replaying sounds.

Sound Waves are Detected by **Diaphragms** in **Microphones**

1) The vibrations in a sound wave make a sensitive diaphragm (e.g. a thin paper or plastic sheet) vibrate inside the microphone.
2) The microphone converts the vibrations to electric signals.
3) Another device can record the electrical signals so that the sound can be reproduced later.

Loudspeakers Recreate Sound Waves

1) An electrical signal is fed into a loudspeaker.
2) This signal causes the diaphragm to vibrate.
3) This makes the air vibrate, producing sound waves.

It's a bit like a microphone in reverse.

Tssk

Tssk

Diaphragm makes air vibrate

Electrical pulses cause vibrations

Ultrasound is **High Frequency** Sound That We **Can't Hear**

1) Ultrasound includes all sounds that have a higher pitch than the normal auditory range of humans.
2) So that'll be any sound over 20 000 Hz.

Ultrasonic Cleaning Uses Ultrasound

High-frequency sound waves are used to clean things — the vibrations of the pressure waves dislodge dirt in tiny cracks that wouldn't normally be cleaned.

1) An item is placed in a special bath filled with water (or another liquid).
2) High-pressure ultrasound waves cause bubbles to form in cavities (holes).
3) The bubbles knock any bits of dirt (contaminants) off the object, leaving it clean enough to eat your dinner off.

You can use ultrasonic cleaning to clean jewellery, false teeth, fountain pen nibs, etc.

Ultrasound **Physiotherapy** May be **Helpful**

1) Ultrasound pressure waves transfer energy through matter — so they can reach inside your body.
2) Some physiotherapists think that this means ultrasound can be used to treat aches and pains in parts of the body that are hard to access — like muscles and tendons deep inside your shoulders.
3) But scientists have found little evidence that ultrasound physiotherapy is an effective treatment.

Warm-Up and Practice Questions

Have a bash at these questions, and see how far you get. If you can't answer some of the questions, or get an answer wrong, go back and read the relevant pages again. Then have another go at these questions until you can do them all.

Warm-Up Questions

1) Which travels faster: light or sound?
2) Name one thing that sound can't travel through.
3) What determines the pitch of sound?
4) Give the lowest frequency humans can typically hear.
5) What can sound waves transfer other than energy?
6) Suggest one material that absorbs sound well.
7) Which part of a microphone vibrates when it detects sound waves?

Practice Questions

1 Sounds are detected by the ear.

(a) What happens to the ear drum when a sound wave hits it?

(1 mark)

(b) Which part of the ear generates electrical signals to send to the brain?

(1 mark)

(c) Some physiotherapists use sounds that cannot be detected by the human ear.

(i) What name is given to this type of sound?

(1 mark)

(ii) Explain why these sounds can't be heard.

(1 mark)

(iii) Give one other use for this type of sound.

(1 mark)

2 Jim is investigating sound waves. He sets up a speaker at one end of his school hall and stands at the opposite end. He sets the speaker to play one loud thump.

(a) Jim feels the thump through his feet just before he hears it. Suggest why.

(1 mark)

(b) After hearing the initial thump, Jim hears a number of other similar but much quieter thumps. Suggest why this might be.

(1 mark)

Revision Summary for Section Ten

Section Ten tells you everything you need to know about waves. There are quite a few words in there — and some pretty important diagrams too. Science isn't always a complete doddle, so you're bound to find some of the facts a bit tricky to learn. Never fear! As somebody famous once said, "Nothing can take the place of persistence" — in other words, if you want to achieve anything worthwhile or difficult, all you have to do is keep on slogging away at it. Better get cracking with this lot then...

1) What do water waves look like? Sketch out a diagram and label it. ☐
2) Describe what happens to the displacement when two waves meet. What is this called? ☐
3) Give three similarities between water waves and light waves. ☐
4) Give one big difference between water waves and light waves. ☐
5) What speed does light travel at in a vacuum? ☐
6) What is meant by a diffuse reflection? ☐
7) What is the law of reflection? ☐
8) What is refraction? ☐
9) What happens when light goes from a less dense medium to a more dense medium? ☐
10) What happens when light goes from a more dense medium to a less dense medium? ☐
11) Explain in your own words why light "bends" as it enters a glass block. ☐
12) Sketch a diagram of a pinhole camera. ☐
13) Use a diagram to explain why the image is upside down and crossed over. ☐
14) Draw a diagram to show how a convex lens refracts parallel rays of light. ☐
15) Which two parts of the eye help you focus on an object? ☐
16) How do digital cameras form images? ☐
17) How could you show that white light is not just one colour? ☐
18) What colour of light has the highest frequency? ☐
19) Why does something red look red in white light? ☐
20) What happens to all the colours in white light when they hit a black object? ☐
21) What colour would green laces look in red light and why? ☐
22) What type of wave are sound (pressure) waves? In which direction are the vibrations? ☐
23) What does sound need to travel from one place to another? ☐
24) Why couldn't you hear a ringing bell in a vacuum? ☐
25) What is an echo? ☐
26) Does sound usually travel faster in solids, liquids or gases? Explain your answer. ☐
27) What does the frequency of a sound mean? ☐
28) Draw a labelled diagram of an ear. Explain how a flicked ruler can be heard. ☐
29) What does auditory range mean? ☐
30) What is the auditory range of humans? ☐
31) How do microphones work? ☐
32) What is ultrasound? What can it be used for? ☐

Electrical Circuits

First up in this section, some electricity basics...

Electric Current is the Flow of Charge

1) Electric current is the flow of charge around a circuit.
2) It can only flow if a circuit is complete.
3) The moving charges are actually negative electrons (page 67).
4) Irritatingly, they flow the opposite way to the direction of conventional current, which is shown on circuits as arrows pointing always from positive to negative.

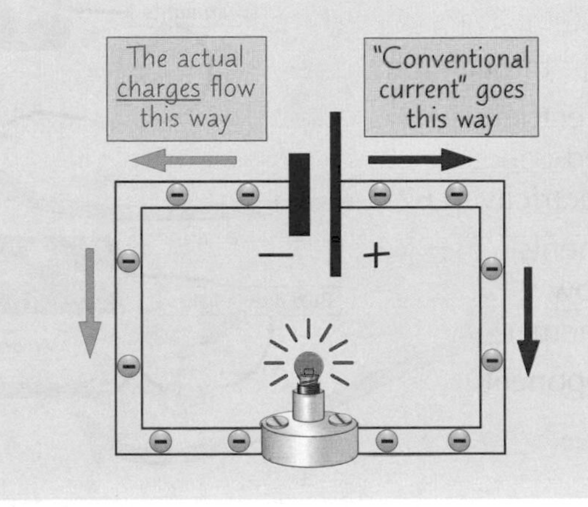

5) It's vital that you realise that CURRENT IS NOT USED UP as it flows through a circuit. The total current in the circuit is always the same.

Current is a bit like Water Flowing

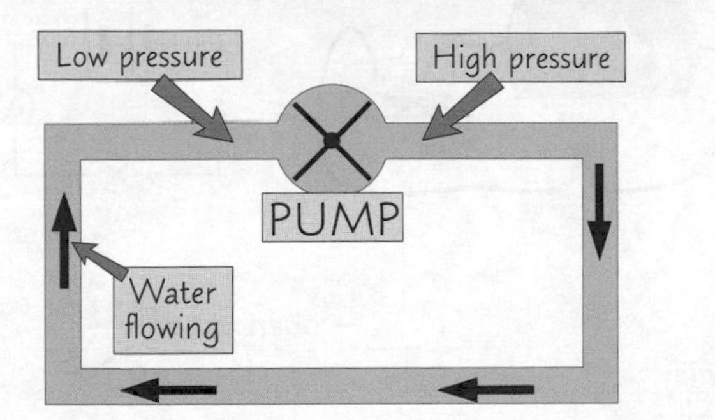

1) The pump drives the water along like a power supply.
2) The water is there at the pump and is still there when it returns to it — and just like the water, electric current in a circuit doesn't get used up either.

Potential Difference Pushes the Current Around

1) In a circuit the battery acts like a pump — it provides the driving force to push the charge round the circuit. This driving force is called the potential difference.
2) If you increase the potential difference more current will flow.
3) Different batteries have different potential differences. You can put several batteries together to make a bigger potential difference too.

Potential difference is sometimes called voltage.

Electrical Circuits

Resistance is How Easily Electricity Can Flow

1) Resistance is anything in a circuit that slows down the flow of current. It is measured in ohms (Ω).
2) You can calculate the resistance of a component by finding the ratio of the potential difference and current. This is just a fancy way of saying:

A component is anything you put in a circuit.

RESISTANCE = POTENTIAL DIFFERENCE ÷ CURRENT

3) This means that as long as the potential difference stays the same, the higher the resistance of a component, the smaller the current through it.
4) Components and materials that electricity can easily travel through are called conductors. Metals are good conductors of electricity (p.67).
5) Insulators (e.g. wood) are components and materials that don't easily allow electric charges to pass through them.
6) The lower the resistance of a component, the better it is at conducting electricity.

Bulb lights

Metal strip

Bulb doesn't light

Wooden or plastic ruler

E.g. bulb A has a resistance of 3 Ω and bulb B has a resistance of 1.5 Ω. Bulb B has a lower resistance, so bulb B is a better conductor than bulb A.

Circuit Diagrams Represent Real Circuits

Circuit diagrams are simplified drawings of real circuits. You start at the cell or battery and go round the circuit, putting in the symbol for each component.

Here are the circuit symbols you need to know:

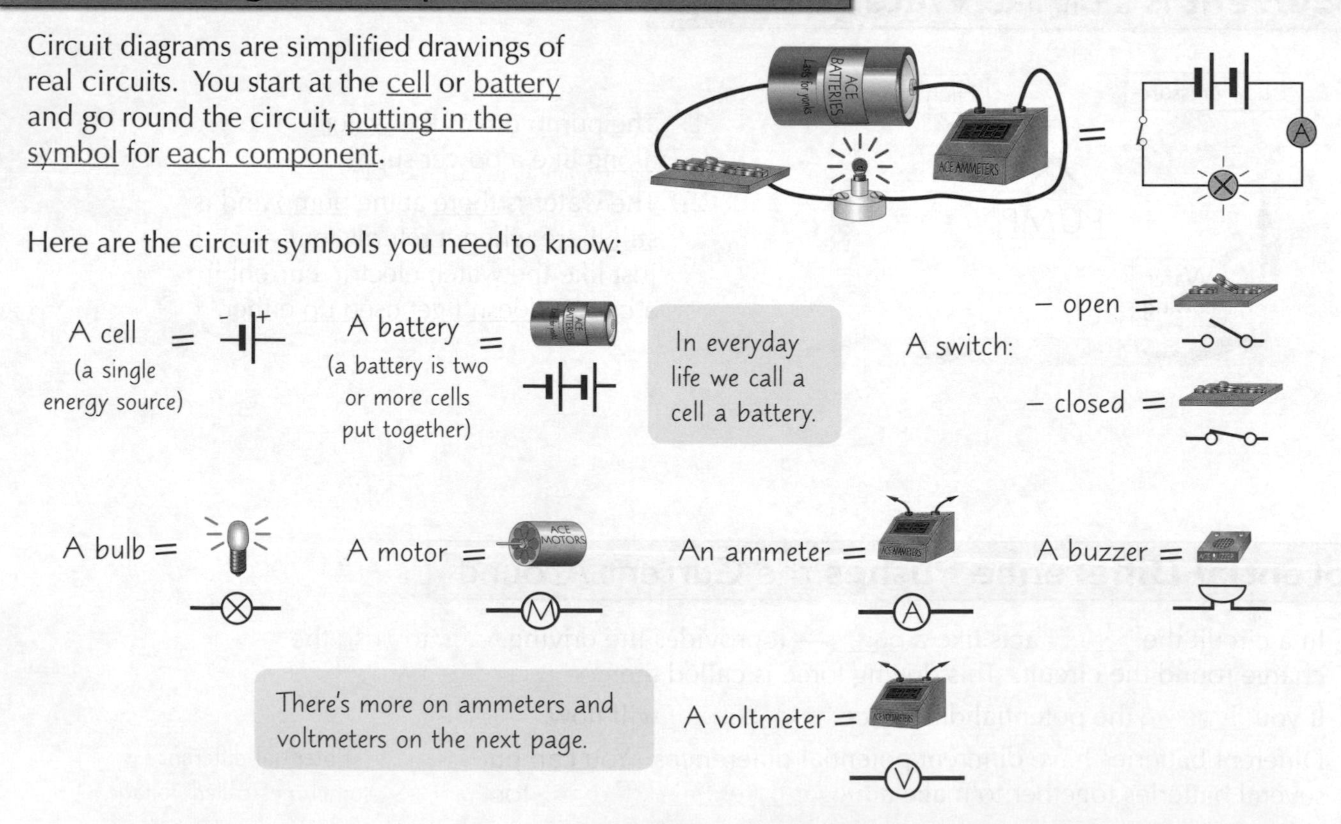

There's more on ammeters and voltmeters on the next page.

Measuring Current and Potential Difference

By now you should know what current and potential difference are (see page 151 for a recap). You need to be able to measure them too. Handily, there are machines to do just that...

Ammeters Measure Current

1) Ammeters measure electric current. It's measured in amperes (or amps, A, for short).
2) You measure the current through a circuit by inserting the ammeter into the circuit like this:

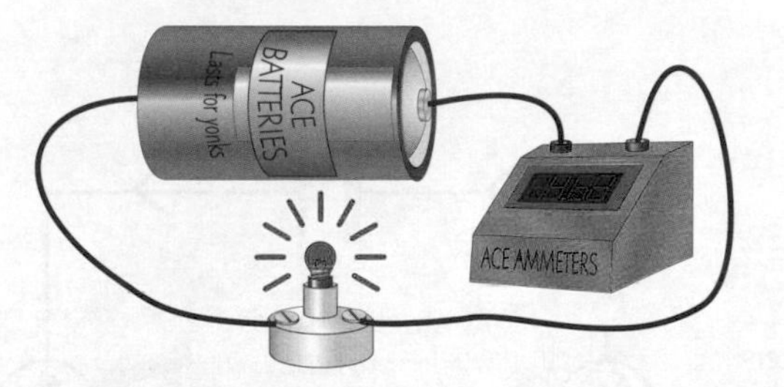

3) Remember — current doesn't get used up, so the current through the ammeter is the same as through the bulb.

Voltmeters Measure Potential Difference

1) Voltmeters measure potential difference in volts (or V for short).
2) You measure the potential difference across something in the circuit, such as a bulb.
3) To measure the potential difference across a bulb, you'd connect a voltmeter across it like this:

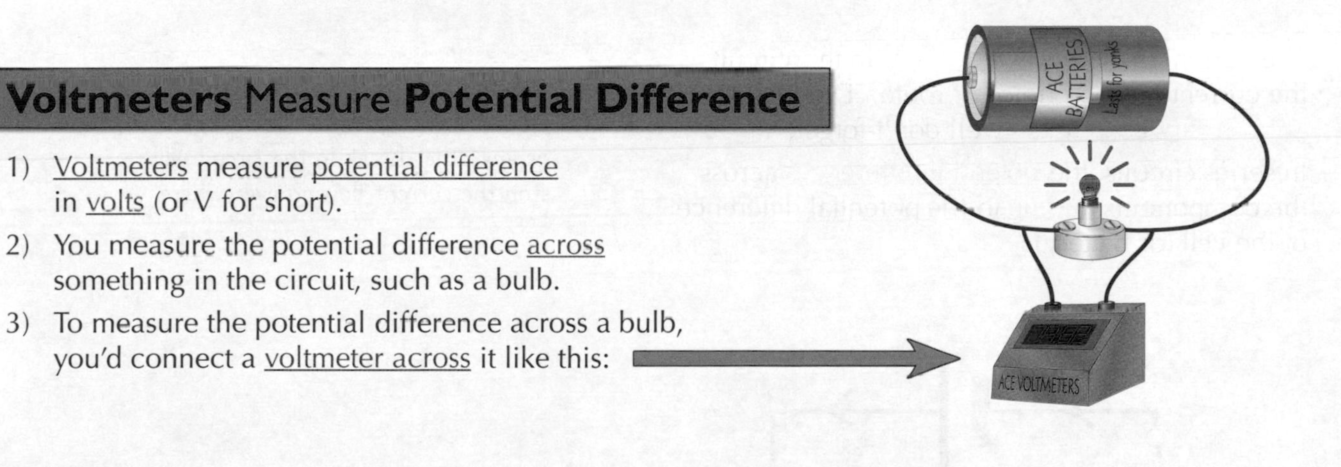

Batteries and Bulbs Have Potential Difference Ratings

1) A battery potential difference rating tells you the potential difference it will supply.
2) A bulb rating tells you the maximum potential difference that you can safely put across it.

Ammeters measure amps, voltmeters measure volts

The stuff on this page should be straightforward — just don't get the two types of meter mixed up.

Series Circuits

In a series circuit, every part of the circuit is on the same path.

Series Circuits — Current has No Choice of Route

1) In the circuit below the current flows out of the cell, through the ammeter and the bulbs, then through the other ammeter and the switch and back to the cell. As it passes through, the current gives up some of its energy to the bulbs.

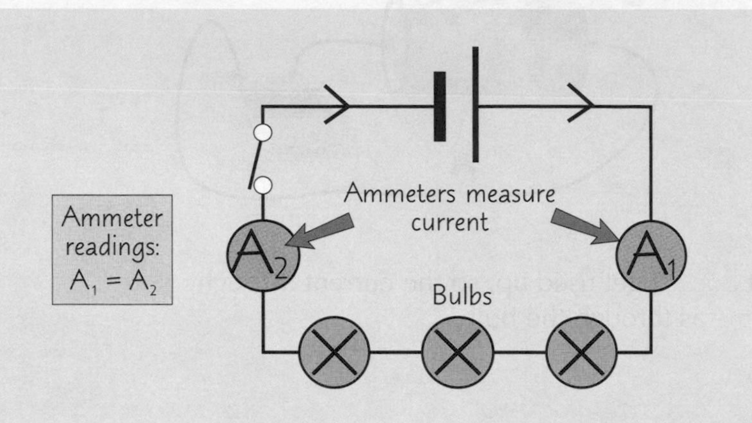

2) The current is the same anywhere in this circuit as the current has no choice of route. Did I tell you current isn't used up — well don't forget.
3) In series circuits, the potential differences across the components add up to the potential difference of the cell (or battery).

In series circuits the current is either on or off — the switch being open or any other break in the circuit will stop the current flowing everywhere.

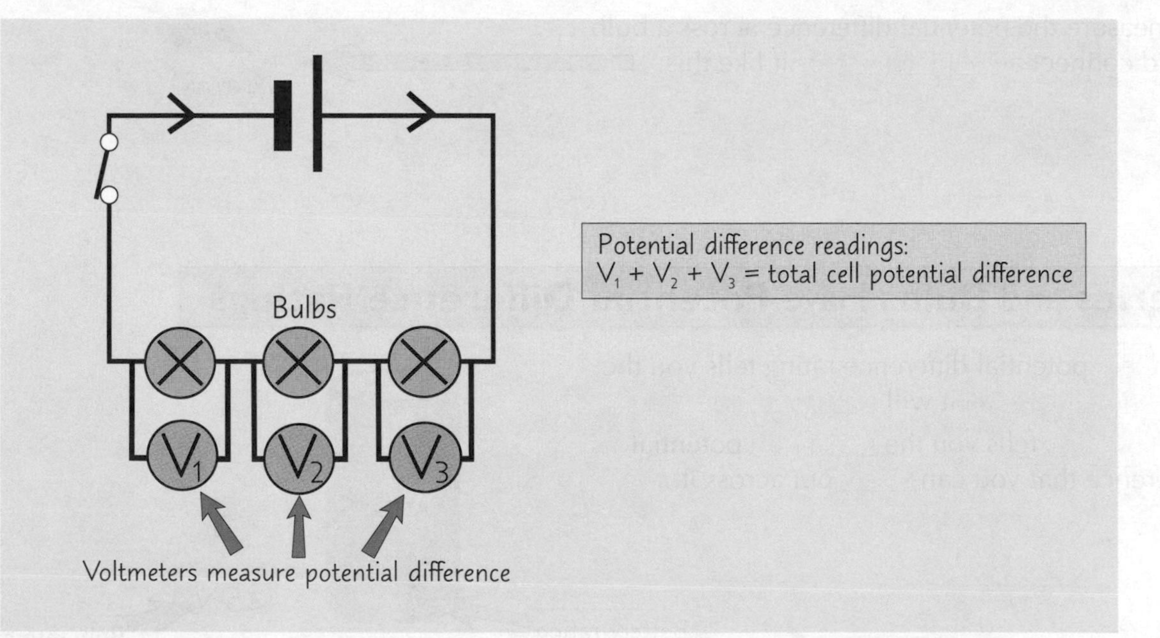

REVISION TIP

Why don't you wire your house lights in series?

If a bulb went, all the lights in a series circuit would go off, leaving you in total darkness. If you're struggling with series circuits, try using the water analogy from page 151. The same 'water' flows through everything, so the current through all of the components is the same.

Parallel Circuits

Parallel circuits are more useful than series circuits in pretty much any situation.

Parallel Circuits — Current has a Choice

1) In the circuit shown below, current flows out of the cell and it all flows through the first ammeter A_1. It then has a "choice" of three routes and the current splits down routes 1, 2 and 3.
2) The readings of ammeters A_3, A_4 and A_5 will usually be different, depending on the resistances of the components — i.e. the bulbs.
3) The three currents join up again on their way back to the cell. So the readings of $A_3+A_4+A_5$ added together will be equal to the reading for current on ammeter A_2 (which will also equal A_1).
4) It's difficult to believe I know, but the current through A_1 is the same as the current through A_2 — the current is NOT USED UP. (I may have told you that once or twice already.)

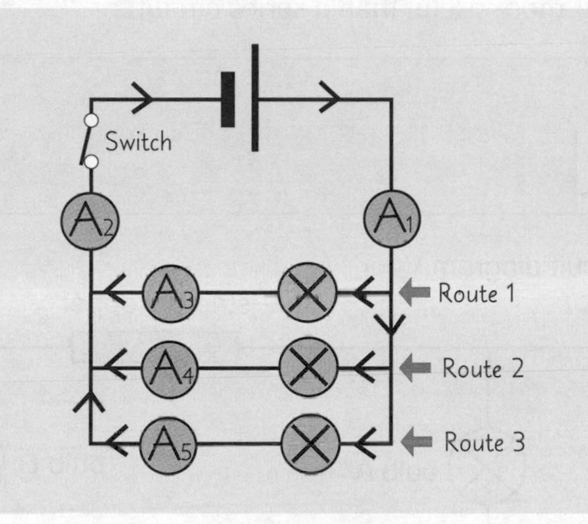

5) Parallel circuits are sensible because part of them can be ON while other bits are OFF. In the circuit below, two bulbs are on and the other one is off.
6) Don't get confused — the potential difference across each bulb in this circuit is equal to the potential difference of the cell.

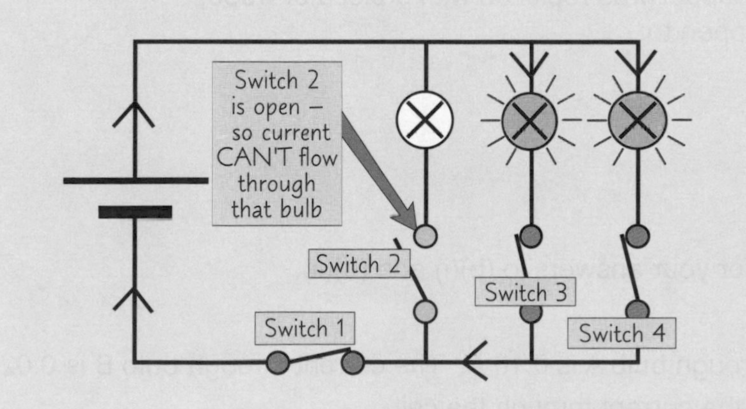

So now you know the two different types of circuits
Make sure you know which is which — it's pretty vital knowledge. Cover up this page and see how much you can scribble down about parallel circuits, then take a break.

Warm-Up and Practice Questions

There's a good mix of Warm-Up Questions to get yourself started. Then you can launch yourself into the more difficult circuit diagram stuff. What a great way to spend an evening.

Warm-Up Questions

1) What is the driving force that pushes current round a circuit?
2) What units is resistance measured in?
3) What is an insulator? Give an example.
4) Draw the circuit symbols for a) a cell b) a bulb c) an ammeter.
5) What do ammeters and voltmeters measure?
6) What is a battery rating? And what is a bulb rating?
7) Why is a parallel circuit more useful than a series circuit?

Practice Question

1 Look at the following circuit diagram.

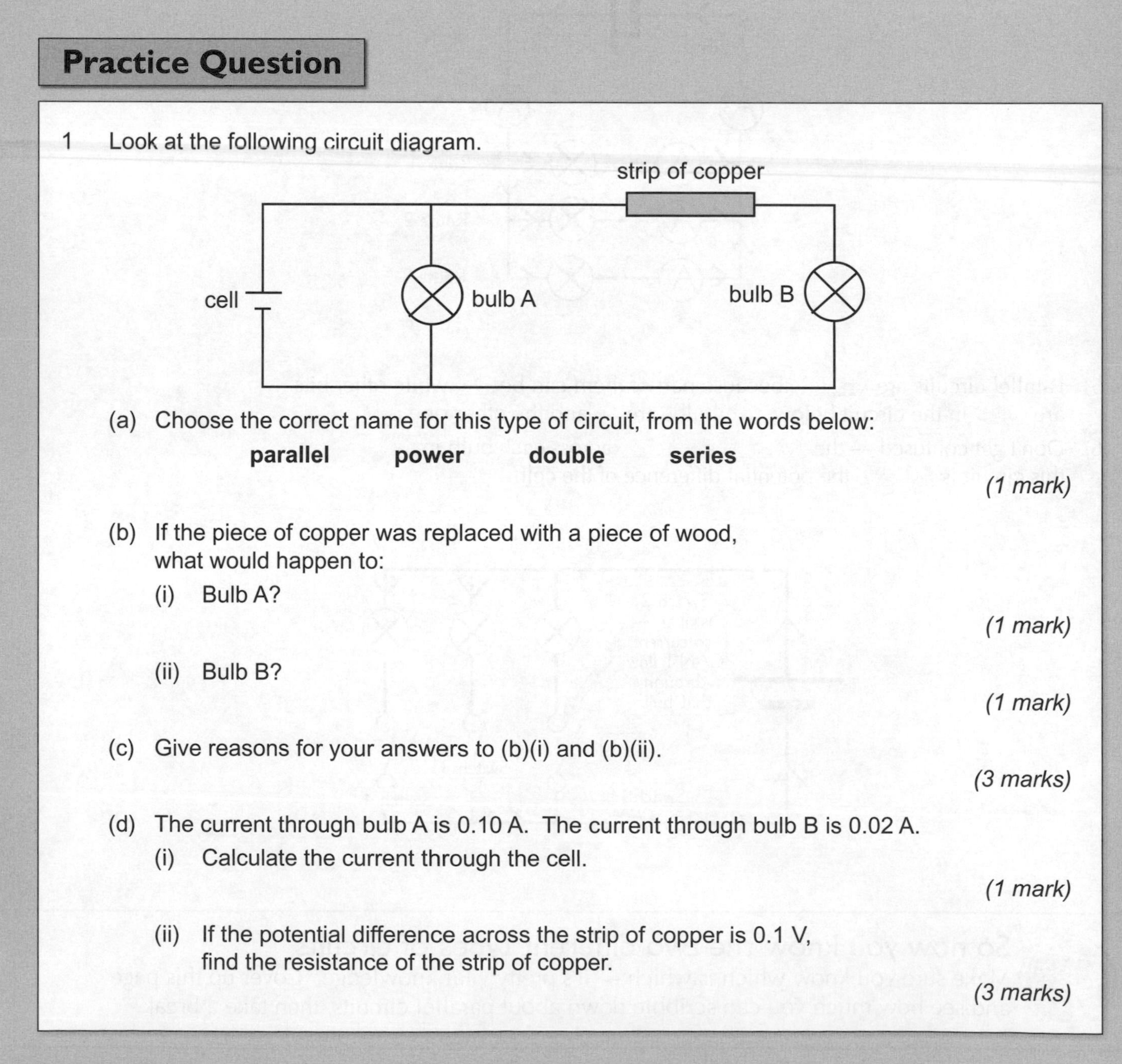

(a) Choose the correct name for this type of circuit, from the words below:

parallel **power** **double** **series**

(1 mark)

(b) If the piece of copper was replaced with a piece of wood, what would happen to:

(i) Bulb A?

(1 mark)

(ii) Bulb B?

(1 mark)

(c) Give reasons for your answers to (b)(i) and (b)(ii).

(3 marks)

(d) The current through bulb A is 0.10 A. The current through bulb B is 0.02 A.

(i) Calculate the current through the cell.

(1 mark)

(ii) If the potential difference across the strip of copper is 0.1 V, find the resistance of the strip of copper.

(3 marks)

Static Electricity

Normally charges flow from one place to another. This page is about the ones that stay where they are.

Charges Can Build Up When Objects are Rubbed Together

1) Atoms (see page 55) contain positive and negative charges.
2) The negative charges are called electrons. Electrons can move, but positive charges can't.
3) When two insulating objects are rubbed together, some electrons are scraped off one object and left on the other.

The object that gains electrons becomes negatively charged.
The object that loses electrons is left with an equal but positive charge.

If you rub a plastic rod with a cloth, electrons move from the cloth to the rod.

rub rub rub rub rub rub

electrons

The places where the electrons left the cloth now have a positive charge.

All Charged Objects Have an Electric Field Around Them

1) Charged objects don't have to touch each other for them to feel a force from each other.
2) An electric field is the space around a charged object where other charged objects will feel a force. That's right, electric forces can act across a gap. Clever stuff.
3) The force charged objects feel when they come near each other depends on what type of charge they have.

Two things with OPPOSITE electric charges are ATTRACTED to each other.
Positive and negative charges attract.

positive charge

negative charge

Two things with the SAME electric charge will REPEL each other.

Electrons move around when objects are rubbed together

When materials are rubbed together it's only ever the electrons that move — the positive charges never ever get to go anywhere. Remember that, it's super important.

Magnets

Electric charges aren't the only things to push and pull each other without touching. Magnets do it too.

Magnets are Surrounded by Fields

1) Bar magnets are (surprisingly enough) magnets that are in the shape of a bar. One end of the bar magnet is called the north pole and the other end is called the south pole.
2) All bar magnets have invisible magnetic fields round them.
3) A magnetic field is a region where magnetic materials (e.g. iron) experience a force.
4) You can draw a magnetic field using lines called magnetic field lines. The magnetic field lines always point from the N-pole to the S-pole.
5) You can investigate magnetic fields using a plotting compass.

magnetic field lines

North pole

South pole

The compass will always point from N to S along the field lines wherever it's placed in the field.

The field lines (or "lines of force") always point from NORTH to SOUTH.

Opposite Poles Attract — Like Poles Repel

1) Just like electric charges (see page 157), magnets don't need to touch for there to be a force between them.
2) North poles and south poles are attracted to each other.
3) If you try and bring two of the same type of magnetic pole together, they repel each other.

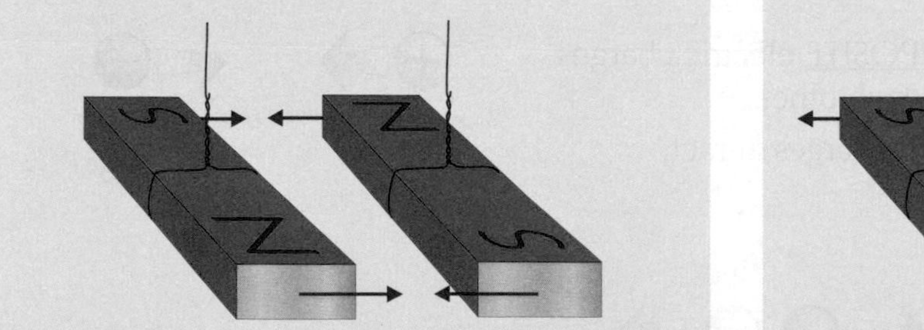

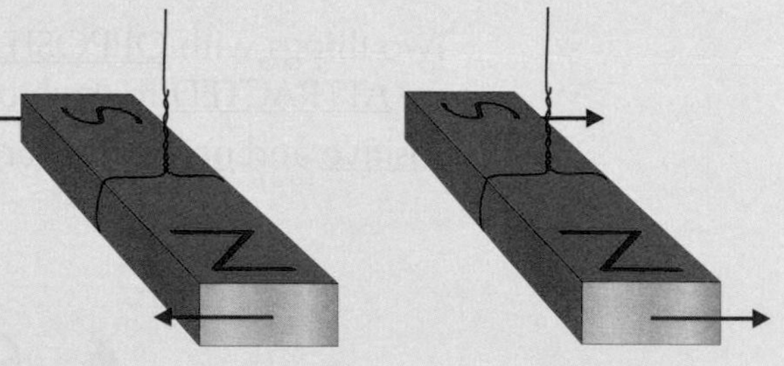

The Earth has a Magnetic Field

1) The Earth has a magnetic field. It has a north pole and a south pole, just like a bar magnet.
2) Compasses line up with magnetic fields — so unless you're stood right next to a magnet, they will point to the Earth's magnetic north pole (which handily is very close to the actual North Pole).
3) Maps always have an arrow on them showing you which direction is north. This means you can use a map and a compass to find your way.

Electromagnets

Bar magnets stay magnetic all the time. Electromagnets can be turned on and off.

A Wire With a Current in it Has a Magnetic Field Round it

1) A current going through a wire causes a magnetic field around the wire.
2) A solenoid is just a long coil of wire. Its magnetic field is the same as that of a bar magnet when a current flows through it.

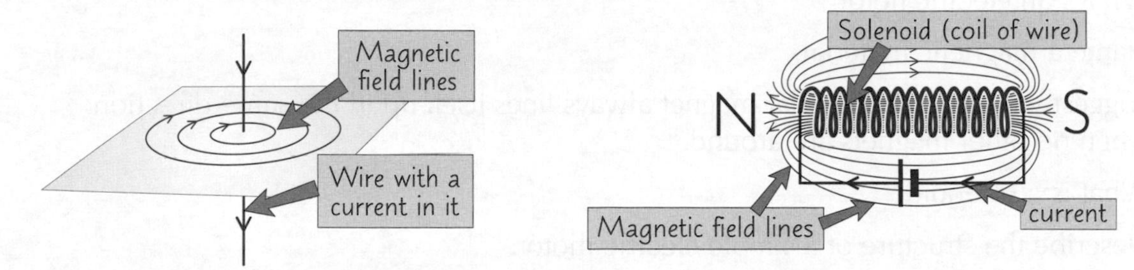

3) Magnets made from a current-carrying wire are called ELECTROMAGNETS.
4) Because you can turn the current on and off, the magnetic field can be turned on and off.

You Can Increase the Strength of an Electromagnet

1) More current in the wire.

Bigger current

2) More turns on the solenoid.

More turns

3) A core of soft iron inside the solenoid.

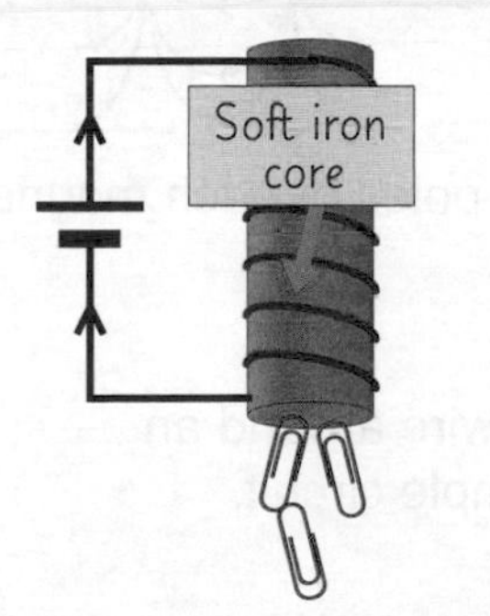

You can't just use any metal to make an electromagnet core. Soft iron has to be used for the core to make it perform as an electromagnet should — i.e. turning on and off when the current is turned on and off.

If a steel core was used, it would stay magnetised after the current was switched off — which would be no good at all.

Electric Motors are Made Using an Electromagnet

1) A simple electric motor is made from a loop of coiled wire in a magnetic field.
2) When current flows through the wire, a magnetic field forms around the wire.
3) Because the wire is already in a magnetic field, there are forces on the loop of the wire. These forces act in opposite directions and cause the loop of wire to turn.
4) Bob's your uncle, you've got a motor.

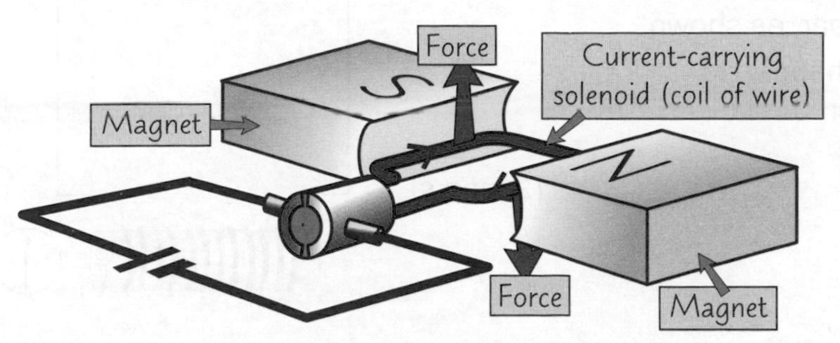

Warm-Up and Practice Questions

The end of another section, and time to test what you've learnt with another set of hand-made Warm-Up and Practice Questions. Why wait — dive straight in and see what you can do.

Warm-Up Questions

1) Static charges are caused by the transfer of which particles?
2) What's an electric field?
3) Name a magnetic material.
4) Suggest why a suspended bar magnet always lines itself up in the same direction when no other magnets are around.
5) What is a solenoid?
6) Describe the structure of a simple electric motor.

Practice Questions

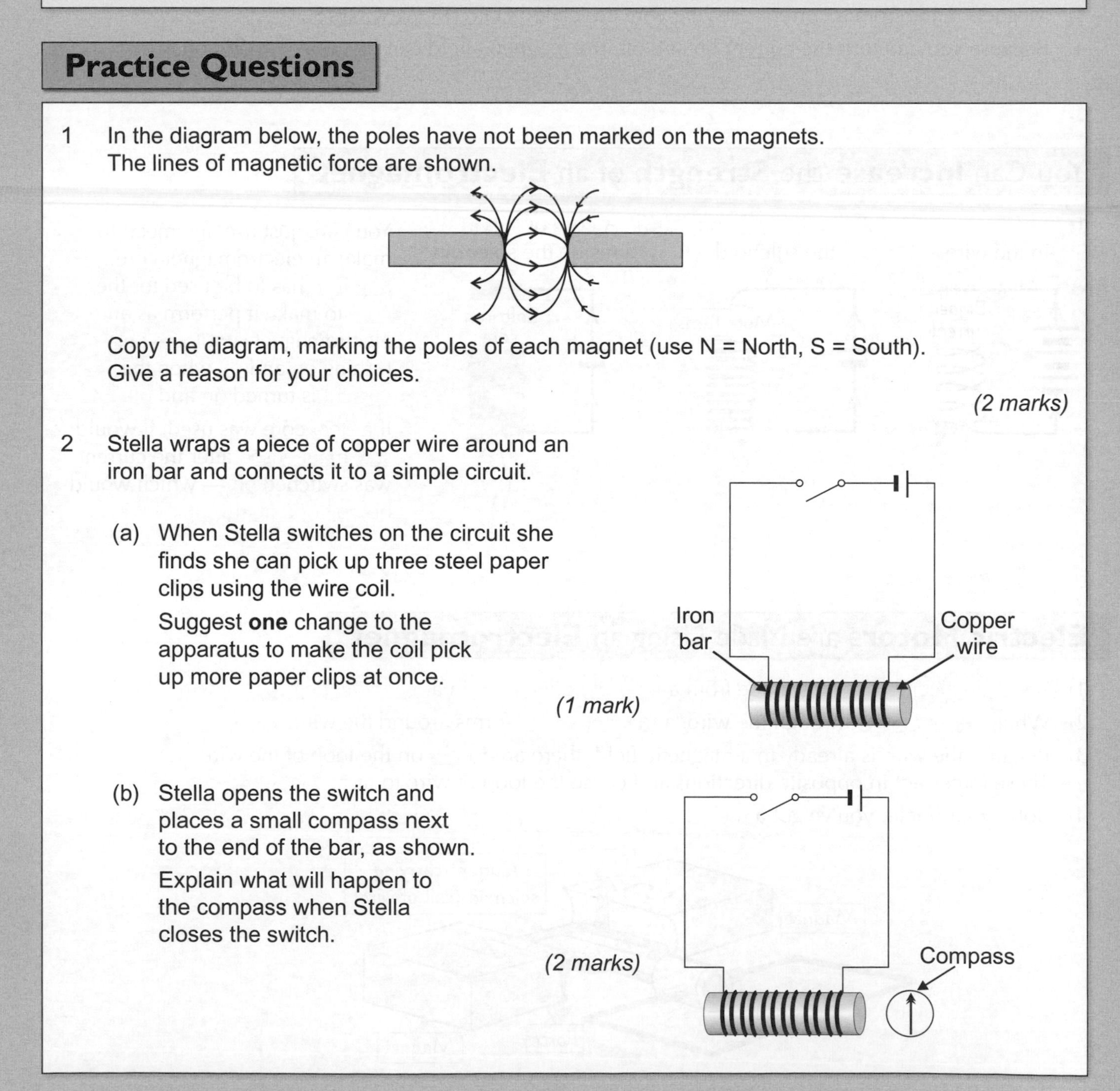

1 In the diagram below, the poles have not been marked on the magnets. The lines of magnetic force are shown.

Copy the diagram, marking the poles of each magnet (use N = North, S = South). Give a reason for your choices.

(2 marks)

2 Stella wraps a piece of copper wire around an iron bar and connects it to a simple circuit.

(a) When Stella switches on the circuit she finds she can pick up three steel paper clips using the wire coil.

Suggest **one** change to the apparatus to make the coil pick up more paper clips at once.

(1 mark)

(b) Stella opens the switch and places a small compass next to the end of the bar, as shown.

Explain what will happen to the compass when Stella closes the switch.

(2 marks)

Revision Summary for Section Eleven

Phew. Electricity and Magnetism — it's no holiday, that's for sure. There are certainly quite a few grisly bits and bobs in this section. There again, life isn't all bad — just look at all these lovely questions I've cooked up for your delight and enjoyment. These are very simple questions which just test how much stuff you've taken on board. They're in the same order as the stuff appears throughout Section Eleven — so for any you can't do, just look back, find the answer, and then learn it good and proper for next time. Yeah that's right, next time — the whole idea of these questions is that you just keep practising them time after time after time — until you can do them all effortlessly.

1) Current is the flow of what?
2) Can current flow in an incomplete circuit?
3) What job does a battery do in a circuit?
4) What is potential difference?
5) What is resistance?
6) What is the difference between a conductor and an insulator?
7)* Component A has a resistance of 1 Ω, Component B has a resistance of 0.5 Ω and Component C has a resistance of 0.01 Ω. Which component is the best electrical conductor?
8) What is a circuit diagram? Why don't we draw out the real thing all the time?
9) Sketch the circuit symbol for all of these:
a) a buzzer b) a battery c) a switch (open) d) a voltmeter.
10) What instrument do we use to measure current? How would you connect it in a circuit?
11) What are the units of current?
12) What instrument do we use to measure potential difference?
How would you connect it in a circuit?
13) What are the units of potential difference?
14)* A series circuit contains 3 bulbs. A current of 3 A flows through the first bulb.
What current flows through the third bulb?
15) What happens if there is a break in a series circuit?
16) Which type of circuit allows part of the circuit to be switched off?
17) In parallel circuits current has a choice of what?
18) True or false? Adding the current through each branch of a parallel circuit gives you the total current.
19) Explain how a cloth and a plastic rod both become charged when they're rubbed together.
20) Do charged objects need to touch to repel each other?
21) State whether each of these pairs of charged objects will be attracted or repelled by each other.
a) positive and positive b) negative and positive c) negative and negative
22) What is a magnetic field? In which direction do field lines always go?
23) Sketch a diagram showing how a plotting compass points around a bar magnet.
24) Name two magnetic poles that will: a) attract each other b) repel each other.
25) Explain how a simple electric motor works.

*Answers on page 198

Gravity

Gravity is a Force that Attracts All Masses

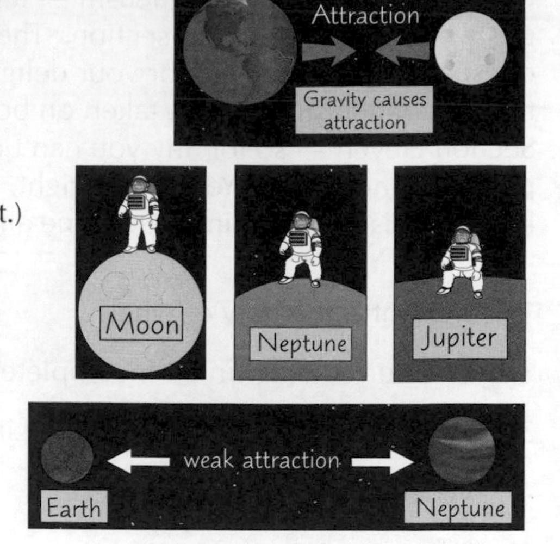

1) Anything with mass will attract anything else with mass. In other words, everything in the Universe is attracted by the force of gravity to everything else.
(But you only notice it when one of the things is really big like a planet.)
2) The Earth and Moon are attracted by gravity — that's what keeps the Moon in its orbit. The Earth and the Sun are attracted by an even bigger force of gravity.
3) The more massive the object (or body) — the stronger the force of gravity is (so planets with a large mass have high gravity).
4) The further the distance between objects — the weaker the gravitational attraction becomes.

Gravity Gives You Weight — But Not Mass

To understand this you must learn all these facts about mass and weight:

1) Mass is just the amount of 'stuff' in an object.
The mass of an object never changes, no matter where it is in the Universe.
2) Weight is caused by the pull of gravity.
3) An object has the same mass whether it's on Earth or on another planet (or on a star) — but its weight will be different. For example, a 1 kg mass will weigh less on Mars (about 3.7 N) than it does on Earth (about 10 N), simply because the force of gravity pulling on it is less.

Weight is a force measured in newtons (N). It's measured using a spring balance or newton meter.
Mass is not a force. It's measured in kilograms (kg) with a mass balance.

Learn this Important Formula...

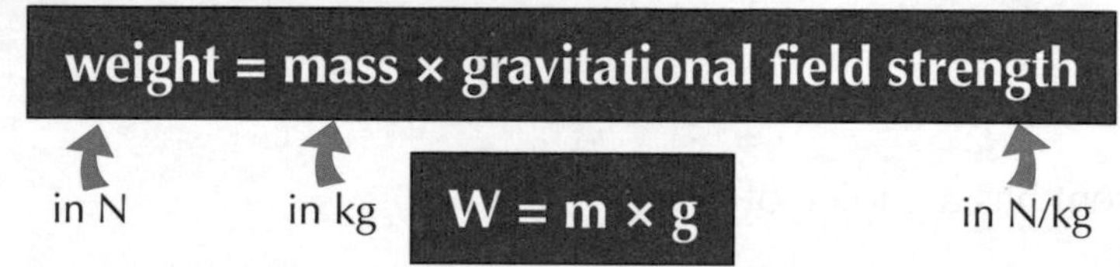

weight = mass × gravitational field strength

in N — in kg — in N/kg

$$W = m \times g$$

1) The letter "g" represents the strength of the gravity and its value is different for different planets. On Earth $g \approx 10$ N/kg. On Mars, where the gravity is weaker, g is only about 3.7 N/kg.
2) This formula is hideously easy to use:

Example: What is the weight, in newtons, of a 5 kg mass, both on Earth and on Mars?

Answer: $W = m \times g$. On Earth: $W = 5 \times 10 = 50$ N (The weight of the 5 kg mass is 50 N.)
On Mars: $W = 5 \times 3.7 = 18.5$ N (The weight of the 5 kg mass is 18.5 N.)

See what I mean? Hideously easy — as long as you've learnt what all the letters mean.

MATHS TIP

Just make sure you appreciate the gravity of all this

If you're asked to do a calculation, make sure you write down the units of your answer. For example, weight is measured in newtons, and mass is measured in kilograms.

The Sun and Stars

You should never, ever look straight at the Sun. But you can look straight at this page, and learn it too.

The **Sun** is at the **Centre** of Our **Solar System**

1) A planet is something which orbits around a star.
2) The Sun (at the centre of our Solar System) is a star. The Earth is one of eight planets which orbit the Sun.
3) The Sun is really huge and has a big mass — so its gravity is really strong. The pull from the Sun's gravity is what keeps all the planets in their orbits.
4) The planets all move in elliptical orbits (elongated circles).
5) Planets don't give out light but the Sun and other stars do.
6) The Sun gives out a massive amount of energy which is transferred by light.

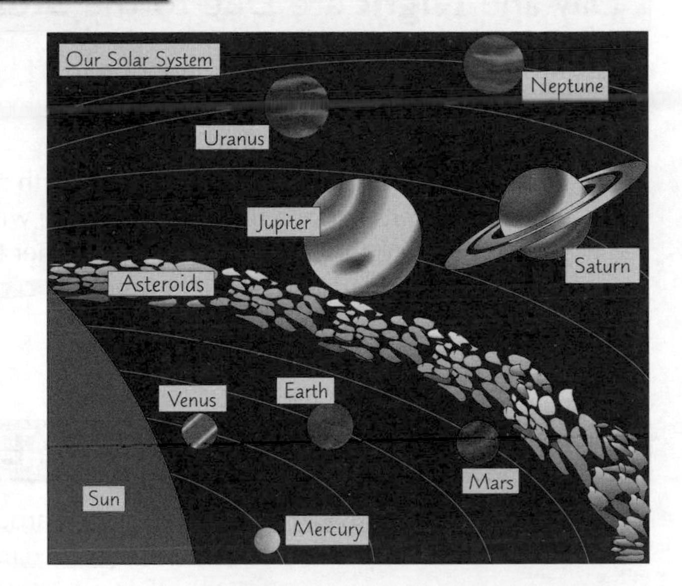

Beyond the **Solar System**

1) A galaxy is a large collection of stars. The Universe is made up of billions of galaxies.
2) Most of the stars you see at night are in our own galaxy — the Milky Way. The other galaxies are all so far away they just look like small fuzzy stars.
3) There are billions of stars in our galaxy, including the Sun.
4) Other stars in our galaxy include the North Star or Pole Star (which appears in the sky above the North Pole) and Proxima Centauri (our nearest star after the Sun).

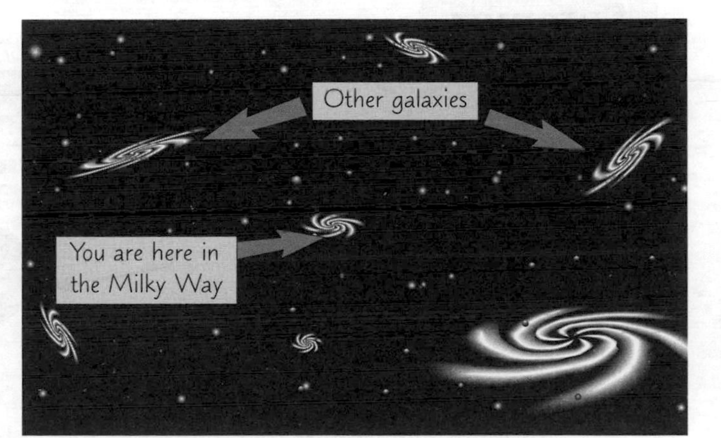

A **Light Year** is a **Unit of Distance**

1) A light year is how far light travels in one year.
2) It's used for measuring huge distances between objects — like the distances you find in space. E.g. Proxima Centauri is about 4 light years away, which means it takes light from the star 4 years to reach Earth.

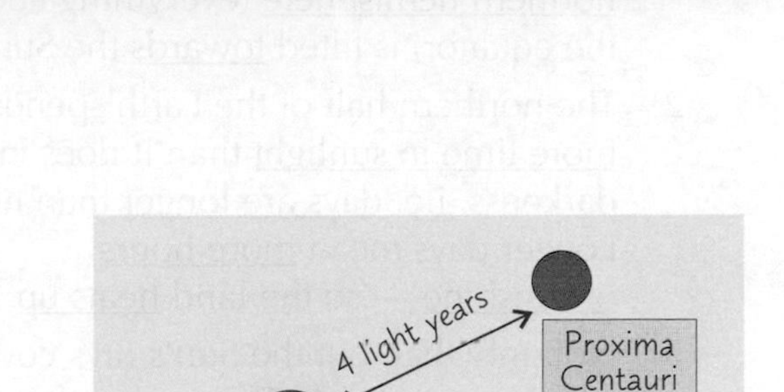

Know the difference between a planet, a star and a galaxy

It sounds obvious, but you really need to get a feel for how these definitions all fit together. You don't have to know the distance to every planet or star, but the more stuff you know the better.

Day and Night and the Four Seasons

In years to come, this stuff will come up in a quiz and you'll be able to wow your teammates with the answer. You also need to know it for KS3 Science. So get cracking...

Day and Night are Due to the Steady Rotation of The Earth

1) The Earth does one complete rotation in 24 hours. That's what a day actually is — one complete rotation of the Earth about its axis.
2) The Sun doesn't really move, so as the Earth rotates, any place on its surface (like England, say) will sometimes face the Sun (day time) and other times face away into dark space (night time).

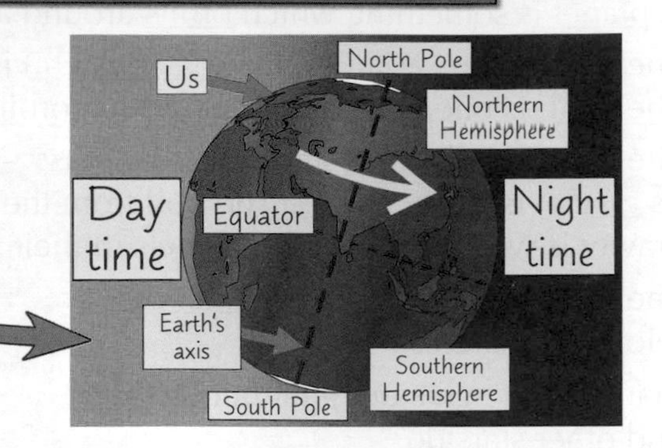

The Seasons are Caused by the Earth's Tilt

1) The Earth takes 365 ¼ days to orbit once around the Sun. That's one year of course. (The extra ¼ day is sorted out every leap year.) Each year has four seasons.
2) The seasons are caused by the tilt of the Earth's axis.

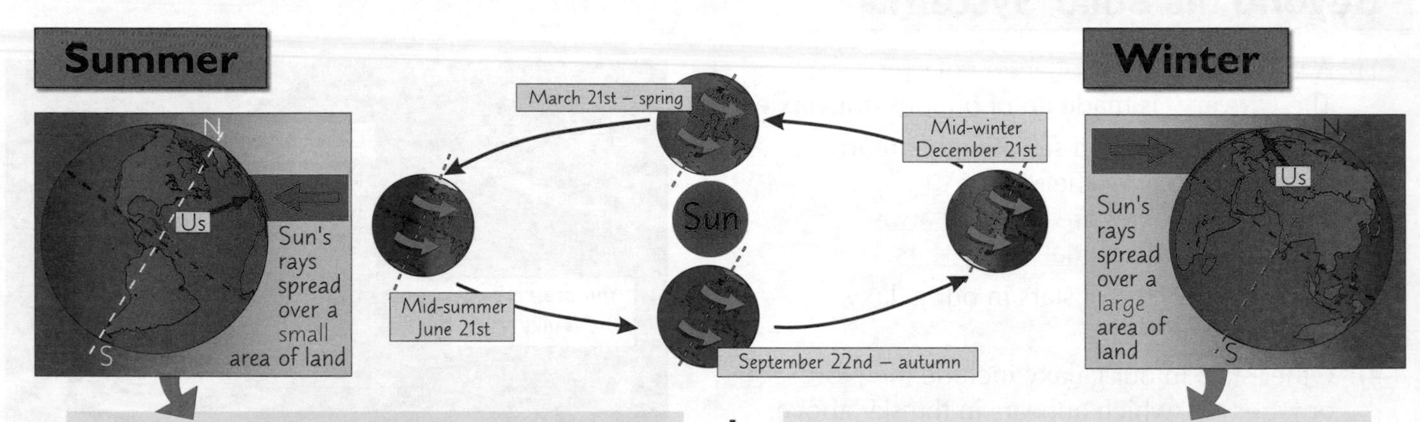

1) When it's summer in the UK, the northern hemisphere (everything above the equator) is tilted towards the Sun.
2) The northern half of the Earth spends more time in sunlight than it does in darkness, i.e. days are longer than nights. Longer days mean more hours of sunshine — so the land heats up...
3) Not only that, but the Sun's rays cover a small area of land. This means that the heat is focused on a small area. So it gets warm and we have summer — hoorah.

1) When it's winter in the UK, the northern hemisphere is tilted away from the Sun.
2) The north now spends less time in sunlight so days are shorter than nights.
3) Also, the Sun's rays cover a larger area of land so the heat is more spread out. So it gets colder and we have winter.

When it's summer in the northern hemisphere, it's winter in the southern hemisphere — and vice versa.

This page is jam-packed with fascinating facts

There's a lot to learn on this page, so you might like to try the mini-essay method. Scribble down a mini-essay that covers all the details on this page. Then check to see what you missed.

Warm-Up and Practice Questions

All this talk about the Universe — I think it's time to actually answer some questions on it.

Warm-Up Questions

1) What's the force of attraction between all bodies called?
2) How many stars are in our Solar System?
3) Suggest why we usually see distant stars more easily than other planets in the Solar System.

Practice Questions

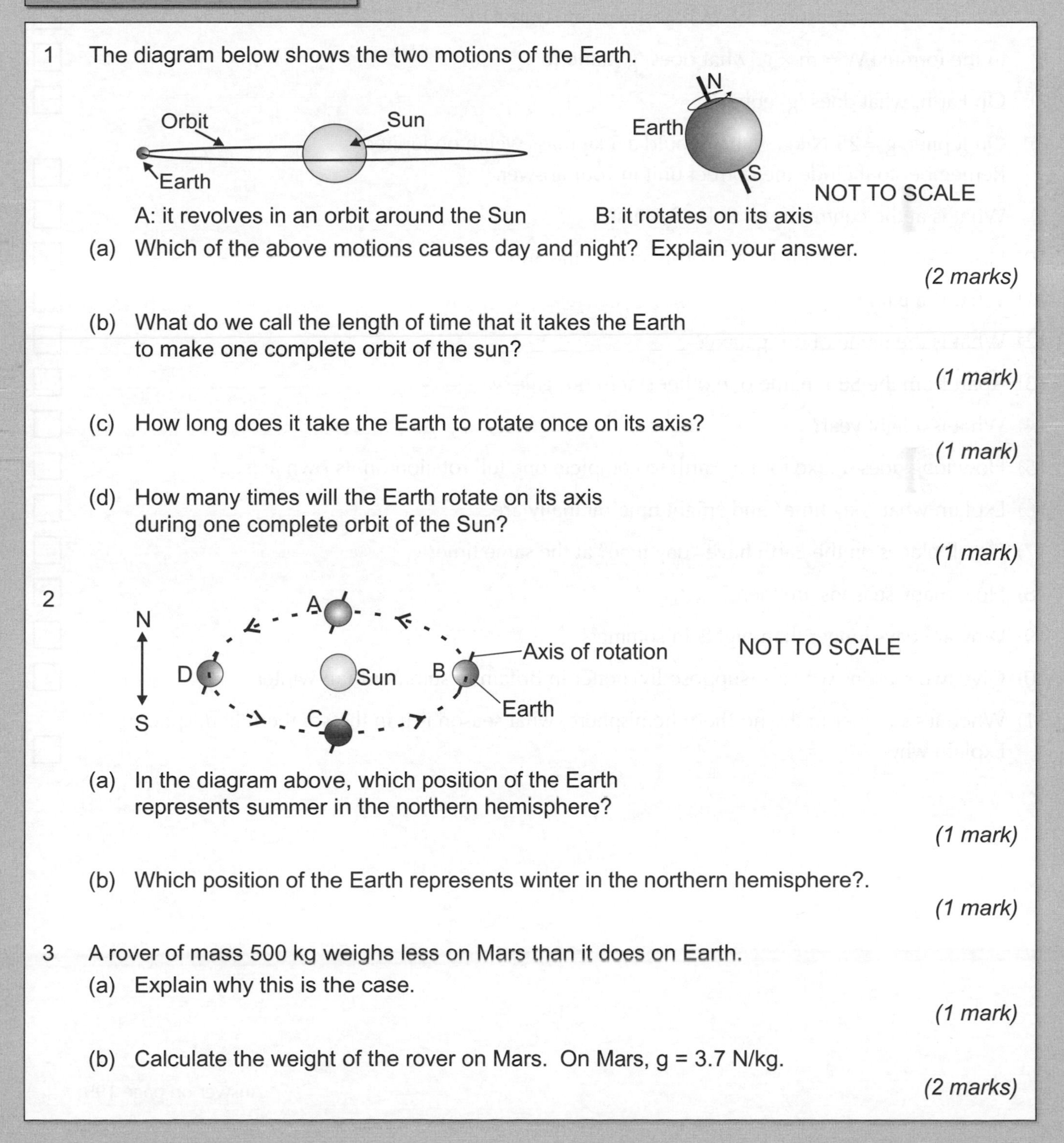

1 The diagram below shows the two motions of the Earth.

A: it revolves in an orbit around the Sun

B: it rotates on its axis

(a) Which of the above motions causes day and night? Explain your answer.

(2 marks)

(b) What do we call the length of time that it takes the Earth to make one complete orbit of the sun?

(1 mark)

(c) How long does it take the Earth to rotate once on its axis?

(1 mark)

(d) How many times will the Earth rotate on its axis during one complete orbit of the Sun?

(1 mark)

2

(a) In the diagram above, which position of the Earth represents summer in the northern hemisphere?

(1 mark)

(b) Which position of the Earth represents winter in the northern hemisphere?.

(1 mark)

3 A rover of mass 500 kg weighs less on Mars than it does on Earth.

(a) Explain why this is the case.

(1 mark)

(b) Calculate the weight of the rover on Mars. On Mars, g = 3.7 N/kg.

(2 marks)

Revision Summary for Section Twelve

Section Twelve only has four pages of information — not much really, considering it deals with the whole Universe. It's amazing just how many people go their whole lives and never really know the answers to all those burning questions, like what is gravity? Or why are the days longer in summer than in winter? Make sure you learn all the burning answers now...

1) What is gravity?
2) Which is stronger, the gravitational attraction between the Moon and the Earth or the Sun and the Earth?
3) How does the mass of a planet affect its gravitational field strength?
4) What is the difference between weight and mass?
5) What is weight measured in? What is mass measured in?
6) In the formula W = m × g, what does 'g' stand for?
7) On Earth, what does 'g' equal?
8)* On Jupiter, g = 25 N/kg. What would a 5 kg mass weigh on Jupiter? Remember to include the correct unit in your answer.
9) What is at the centre of our Solar System?
10) How are all the planets kept in orbit around the Sun?
11) What is a galaxy?
12) What is the name of our galaxy?
13) Apart from the Sun, name one other star in our galaxy.
14) What is a light year?
15) How long does it take for the Earth to complete one full rotation on its own axis?
16) Explain what "day time" and "night time" actually are.
17) Do all places on the Earth have "day time" at the same time?
18) How many seasons are there?
19) Why are days longer than nights in summer?
20) Give two reasons why it's (supposedly) hotter in Britain in summer than winter.
21) When it's summer in the northern hemisphere, what season is it in the southern hemisphere? Explain why.

* Answer on page 199.

Mixed Practice Tests

OK, so you've done most of the hard work — but are you ready for the big Practice Exam?
To help you decide, here are some brilliant Mixed Practice Tests for you to have a go at.

- Scribble down your answers to the questions in a test. When you've answered them all, check your answers (see pages 199-200). Tick the box next to each question you got right. Put a cross in the box if you got it wrong.
- If you're getting 7 or more out of 10 right on these tests, you should be ready for the Practice Exam on page 173.
- If you're getting less than that, go back and do some more revision. Have another go at the Section Summaries — they're the best way to find out what you know and what you've forgotten.

Test 1

✓ / ×

1. What is the pH of a neutral substance?

   A 0
   B 1
   C 7
   D 14

2. Name the energy store that energy is transferred to when an object's height above the ground is increased.

3. What is missing from this sequence showing cell organisation: cell → ...... → organ?

4. Which of the following word equations shows fermentation?

   A glucose + oxygen → carbon dioxide + water + energy
   B glucose + oxygen → lactic acid + energy
   C glucose → lactic acid + energy
   D glucose → carbon dioxide + ethanol + energy

5. State Hooke's law.

6. Name two of the seven life processes.

7. During the summer in the northern hemisphere, ...

   A ...the northern hemisphere is tilted towards the Sun and the Sun's rays are spread over a small area.
   B ...the northern hemisphere is tilted away from the Sun and the Sun's rays are spread over a small area.
   C ...the northern hemisphere is tilted towards the Sun and the Sun's rays are spread over a large area.
   D ... the northern hemisphere is tilted away from the Sun and the Sun's rays are spread over a large area.

8. What's the difference between a physical change and a chemical change?

9. Give two properties of ceramics.

10. What is the crust of the Earth?

Total (out of 10):

Mixed Practice Tests

Test 2

✓ / ×

1. Give two benefits of recycling.

2. Name two ways that energy can be transferred by heating.

3. Which of the following equations could you use to calculate the weight of an object?

   A weight = mass × gravitational field strength
   B weight = mass ÷ gravitational field strength
   C weight = ½ × mass × gravitational field strength
   D weight = gravitational field strength ÷ mass

4. Give an example of a pair of antagonistic muscles.

5. Diffusion is...

   A ...when particles spread from areas of high to low concentration.
   B ...when particles spread from areas of low to high concentration.
   C ...when a liquid and a solid are mixed to form a solution.
   D ...when a liquid turns into a gas.

6. Give four functions of the human skeleton.

7. At approximately what day in the menstrual cycle is an egg released from the ovaries?

   A Day 1
   B Day 4
   C Day 14
   D Day 28

8. How did giraffes end up with very long necks?

   A Generations of giraffes stretched their necks to reach trees.
   B Giraffes with long necks were more likely to survive and reproduce.
   C Giraffes naturally consider long necks as a sexually attractive trait.
   D A spontaneous genetic mutation occurred.

9. Two resistors, X and Y, are connected in parallel to each other, as shown in the diagram.
   The current through resistor X is 5 A.
   The current through resistor Y 10 A.
   What is the total current through the whole circuit?

   X
   Y
   $I = 5$ A
   $I = 10$ A

   A 5 A
   B 10 A
   C 15 A
   D 50 A

10. Balance the following chemical equation: $3\,S + 2\,O_3 \rightarrow ....SO_2$.

Total (out of 10):

Mixed Practice Tests

Test 3

✓ / ✗

1. A battery-powered fan is turned on. Energy is transferred...
   A ...electrically from the electrostatic energy store of the battery to the kinetic energy stores of the motor and fan blades.
   B ...electrically from the chemical energy store of the battery to the kinetic energy stores of the motor and fan blades.
   C ...mechanically from the chemical energy store of the battery to the kinetic energy stores of the motor and fan blades.
   D ...mechanically from the chemical energy store of the battery to the electrostatic energy stores of the motor and fan blades.

2. What happens when a substance sublimates?

3. The amplitude of a wave is...
   A ...the highest part of the wave.
   B ...the maximum displacement of a wave.
   C ...the lowest part of the wave.
   D ...the length of the wave.

4. Otters eat pike, pike eat minnow and minnow eat waterweed.
   Which of them is the primary consumer?
   A Otter
   B Pike
   C Minnow
   D Waterweed

5. The potential difference across a component is 12 V.
   The current through the component is 2 A. What is the resistance of the component?

6. What is thermal decomposition?

7. What does it mean when the forces on an object are said to be 'in equilibrium'?

8. What feature of a plant helps wind pollination?
   A Brightly coloured petals
   B Short filaments
   C Feathery stigma
   D Nectaries

9. Saliva contains an enzyme called amylase. What is an enzyme?

10. The main role of proteins in your diet is...
    A ...to provide energy for your body.
    B ...to aid the growth and repair of cells.
    C ...to help food move through your digestive system.
    D ...to send signals in nerves.

Total (out of 10):

Mixed Practice Tests

Test 4

✓ / ×

1. What is an ecosystem?

2. Particles of a substance in a solid state...
   A ...are held together by strong forces and have less energy than in the liquid state.
   B ...are held together by strong forces and have more energy than in the liquid state.
   C ...are held together by weak forces and have less energy than in the liquid state.
   D ...are held together by weak forces and have more energy than in the liquid state.

3. Give one difference between the typical properties of metals and non-metals.

4. What is an electromagnet?

5. Which of the following equations could you use to calculate the speed of an object?
   A speed = distance × time
   B speed = distance ÷ time
   C speed = time ÷ distance
   D speed = 2 × distance × time

6. What is the speed of light in a vacuum?

7. Which of the following is the correct word equation for photosynthesis?
   A carbon dioxide + water → glucose + oxygen
   B glucose + water → carbon dioxide + oxygen
   C oxygen + water → glucose + carbon dioxide
   D carbon dioxide + glucose → water + oxygen

8. Some characteristics, like eye colour, are hereditary.
   What does the term 'hereditary' mean?

9. The current composition of the atmosphere is approximately...
   A 78% nitrogen, 21% oxygen, 0.04% carbon dioxide
   B 78% oxygen, 21% nitrogen, 0.04% carbon dioxide
   C 75% oxygen, 21% nitrogen, 4% carbon dioxide
   D 75% nitrogen, 21% oxygen, 4% carbon dioxide

10. Which of the following word equations correctly shows what occurs when an acid and an alkali are mixed?
   A acid + alkali → salt + hydrogen
   B acid + alkali → metal oxide + water
   C acid + alkali → hydrogen + water
   D acid + alkali → salt + water

Total (out of 10):

Mixed Practice Tests

Test 5

✓ / ×

1. What is an omnivore?

2. A substance contains atoms of different types, joined together. The substance is...

   A ...an element.
   B ...a molecule.
   C ...a compound.
   D ...a mixture.

3. Which of the following equations could you use to calculate the daily basic energy requirement of a person?

   A daily BER = 24 hours × body mass
   B daily BER = 5.4 × 24 hours × body mass
   C daily BER = (24 hours × body mass) ÷ 5.4
   D daily BER = 5.4 × body mass

4. Put the following in order of reactivity, from most to least reactive: magnesium, copper, potassium, zinc.

5. What is the typical auditory range of human hearing?

6. The gradient of a distance-time graph shows...

   A ...the total distance travelled by the object.
   B ...the time an object has been travelling.
   C ...the speed the object is travelling at.
   D ...the acceleration of the object.

7. What is meant by 'discontinuous variation'?

8. A 40 W radio is left on for two minutes. How much energy is transferred by the radio in this time?

   A 20 J
   B 80 J
   C 480 J
   D 4800 J

9. A plastic rod is rubbed with a cloth. The plastic rod becomes negatively charged. Which of the following describes what has happened?

   A Negative charges have moved from the rod to the cloth.
   B Negative charges have moved from the cloth to the rod.
   C Positive charges have moved from the rod to the cloth.
   D Positive charges have moved from the cloth to the rod.

10. Calcium and oxygen are combined to make the compound CaO. What is the name of this compound?

Total (out of 10):

Mixed Practice Tests

Test 6

✓ / ×

1. What colour does universal indicator turn when mixed with a strong acid?

   A Red
   B Yellow
   C Green
   D Purple

2. What is a gamete?

3. The mitochondria of a cell...

   A ...are filled with cell sap.
   B ...are where most of the aerobic respiration takes place.
   C ...controls what happens in the cell.
   D ...controls what passes in and out of the cell.

4. Is energy transferred to or taken in from the surroundings during an endothermic reaction?

5. Which of the following equations could you use to calculate pressure?

   A pressure = force × area
   B pressure = force^2 × area
   C pressure = force ÷ area
   D pressure = area ÷ force

6. The upthrust on an object is less than its weight. Will the object float?

7. Which of the following equations could you use to calculate the moment of a force?

   A moment = force ÷ perpendicular distance
   B moment = perpendicular distance ÷ force
   C moment = 2 × force × perpendicular distance
   D moment = force × perpendicular distance

8. Give two ways that alcohol has a negative effect on the human body.

9. Metamorphic rocks...

   A ...form from magma that cools quickly above ground.
   B ...form from magma that cools slowly underground.
   C ...form from layers of sediment.
   D ...form from existing rocks that are exposed to heat and pressure.

10. Name two processes in the carbon cycle that release CO_2 into the air.

Total (out of 10):

CGP

Key Stage 3 Science Practice Exam

Instructions

- The test is one hour long.
- Make sure you have these things with you before you start: pen, pencil, rubber, ruler, angle measurer or protractor, pair of compasses and a calculator.
- The easier questions are at the start of the test.
- Try to answer all of the questions.
- Don't use any rough paper — write all your answers and working in this test paper.
- Check your work carefully before the end of the test.

Score: ☐ **out of 75**

---

1. The diagram below shows an outline of the Periodic Table. Four of the elements have been labelled.

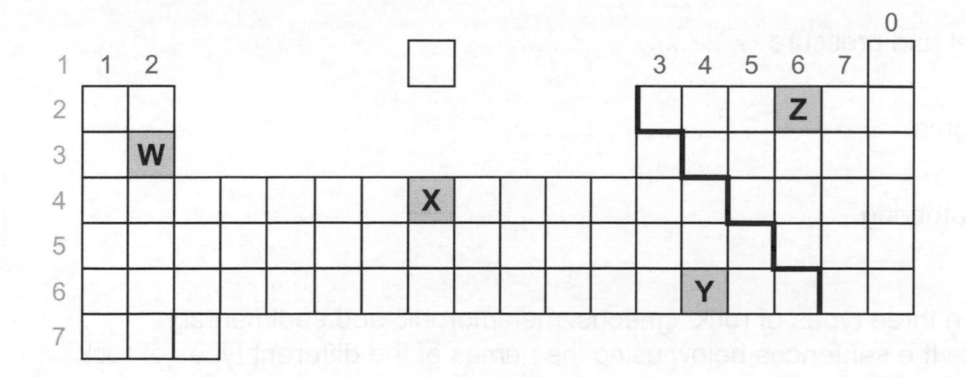

(a) Which of the labelled elements, **W**, **X**, **Y** or **Z**, is in **Group 2**?

....

1 mark

(b) Elements in the Periodic Table can be metals or non-metals.

(i) Which of the labelled elements, **W**, **X**, **Y** or **Z**, is a **non-metal**?

....

1 mark

(ii) Verity has a sample of an element that is a metal.
Give **two** properties that you would expect Verity's sample to have.

1. ....

2. ....

2 marks

2. The diagram below shows the rock cycle.

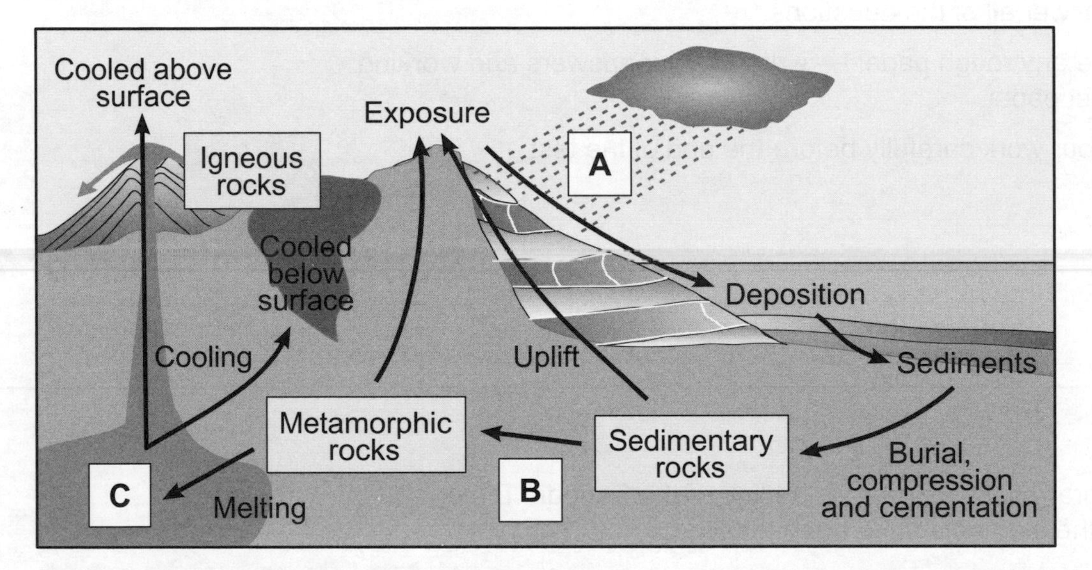

(a) Write the correct letter from the diagram next to each of the labels below.

(i) Heat and pressure ....

(ii) Magma ....

(iii) Weathering ....

3 marks

(b) There are three types of rock: igneous, metamorphic and sedimentary.
Complete the sentences below using the names of the different types of rock.

(i) .... rocks are formed from layers of rock fragments and dead matter, laid down and squashed together over many years.

(ii) .... rocks have no layers. They are made up of lots of crystals, which may be large or small.

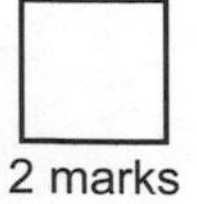

2 marks

3. All living organisms are made up of cells.

(a) The diagram below shows a sketch of a typical animal cell.
Some of the labels are missing.

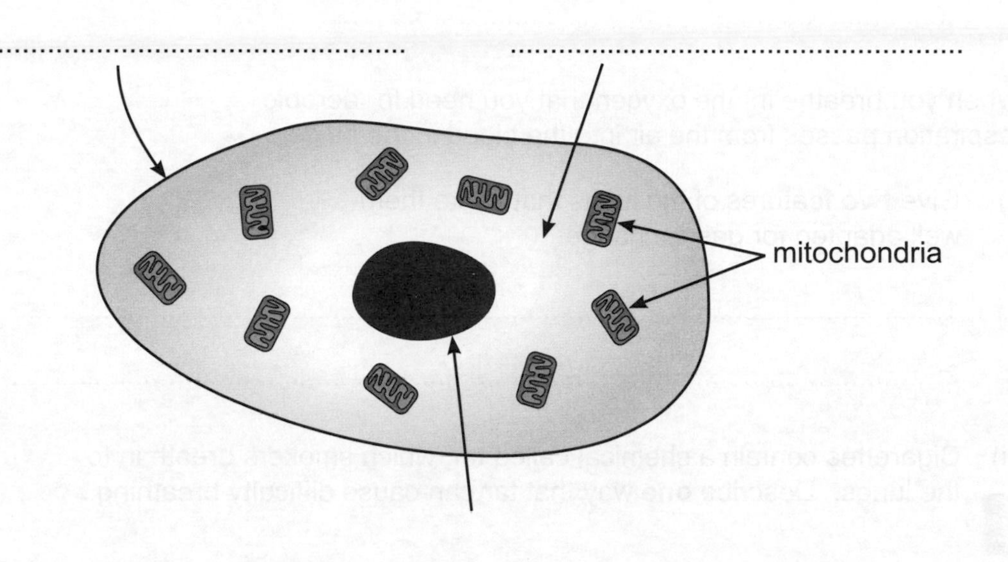

(i) Complete the diagram by filling in the missing labels.

3 marks

(ii) Give the function of the mitochondria.

....

....

1 mark

(b) *Chlamydomonas reinhardtii* is a single-celled organism that lives in water.
Each cell has two tail-like structures, called flagella.

Suggest why *Chlamydomonas reinhardtii* cells have flagella.

....

1 mark

4. Respiration takes place in every cell of every living organism. The main type of respiration that takes place inside human body cells is aerobic respiration.

(a) Complete the word equation below for aerobic respiration.

.... + oxygen → .... + water (+ energy)

1 mark

(b) When you breathe in, the oxygen that you need for aerobic respiration passes from the air into the blood in the lungs.

(i) Give **two** features of the lungs that make them well adapted for gas exchange.

1. ....

2. ....

2 marks

(ii) Cigarettes contain a chemical called tar, which smokers breath in to the lungs. Describe **one** way that tar can cause difficulty breathing.

....

....

1 mark

(c) Another type of respiration that can take place in human cells is anaerobic respiration. What is the main way that anaerobic respiration is difference to aerobic respiration?

....

....

1 mark

5. Jack and Rico are using a glass cell containing smoke to investigate smoke particles using the apparatus shown below.

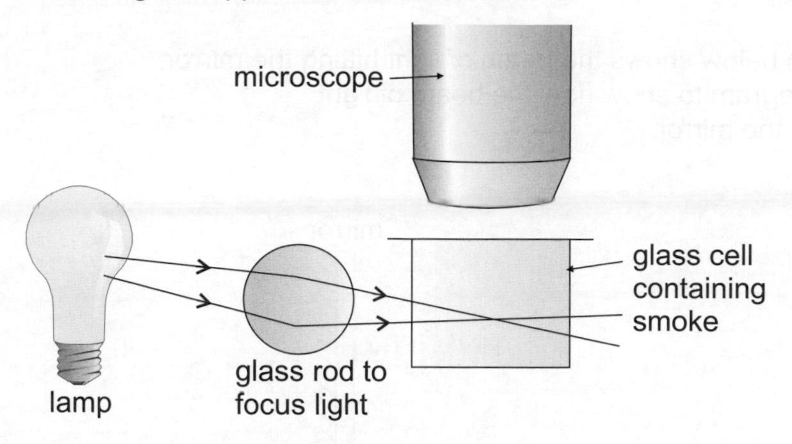

(a) Jack observes the movement of the smoke particles in the cell using the microscope.

(i) Explain why the smoke particles move around randomly.

....

.... 1 mark

(ii) Name the type of movement shown by the smoke particles.

.... 1 mark

(b) Rico accidentally releases the smoke from the glass cell.
Just after this happens, Jack cannot smell the smoke a few metres away.
After 30 seconds pass, Jack can smell the smoke.
Explain why.

....

....

.... 2 marks

6. Amber is investigating how light is reflected.
She shines a beam of light at a mirror.

(a) The ray diagram below shows the beam of light hitting the mirror.
Complete the diagram to show how the beam of light is reflected from the mirror.

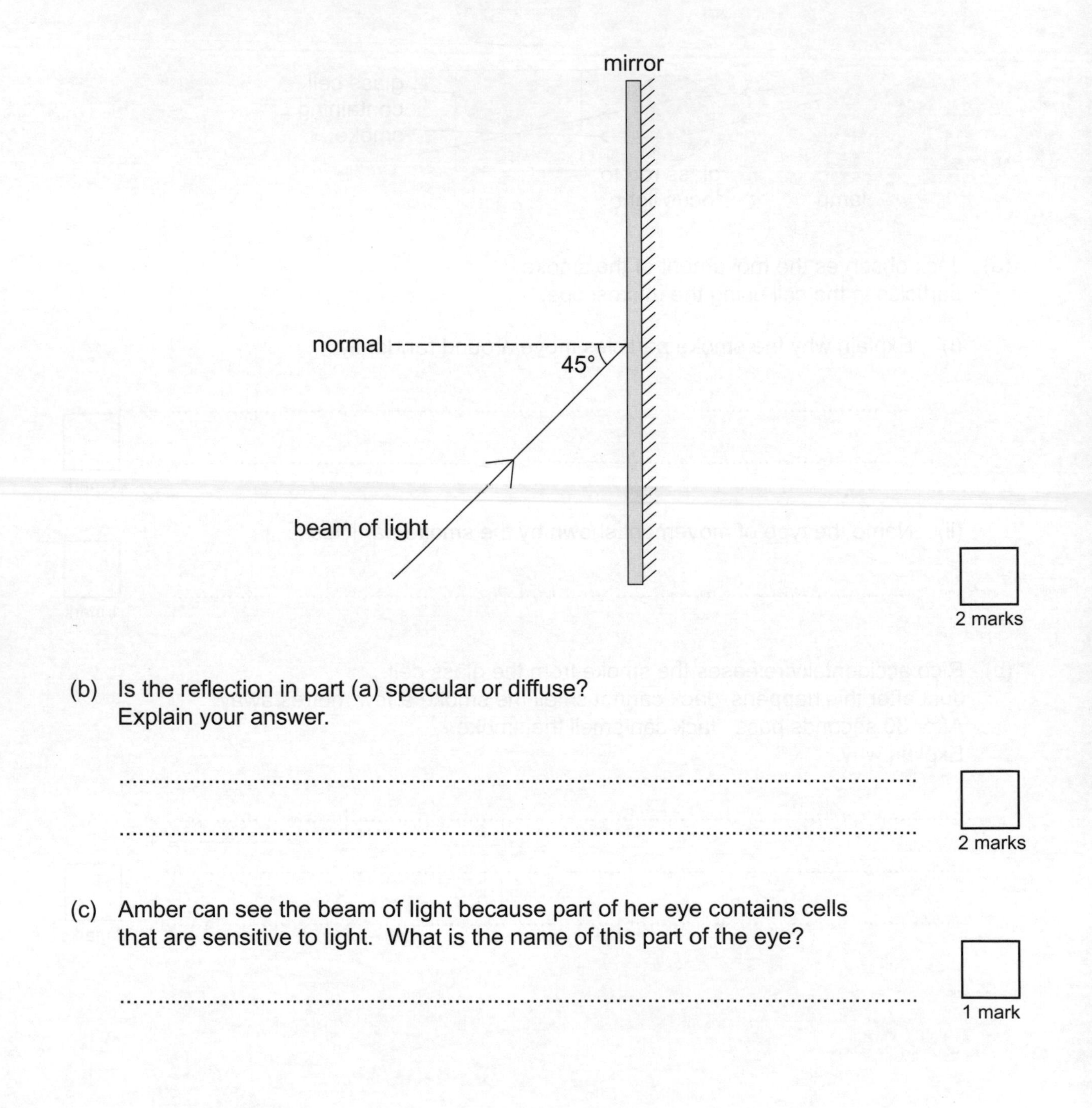

2 marks

(b) Is the reflection in part (a) specular or diffuse?
Explain your answer.

....

....

2 marks

(c) Amber can see the beam of light because part of her eye contains cells that are sensitive to light. What is the name of this part of the eye?

....

1 mark

7. Sati is making her own ink by dissolving blue dye in water.

(a) Identify the solvent, solution and solute in Sati's mixture.

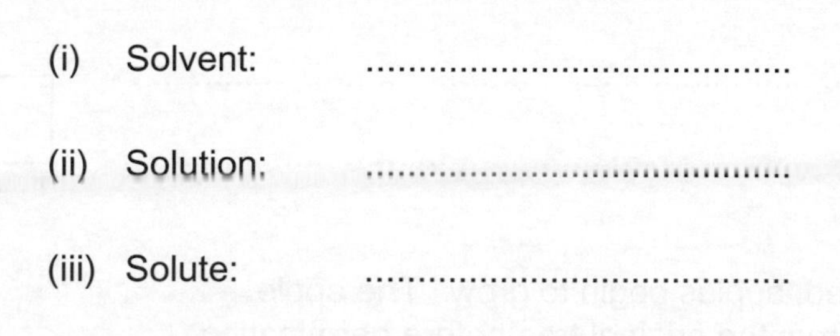

(i) Solvent: ....

(ii) Solution: ....

(iii) Solute: ....

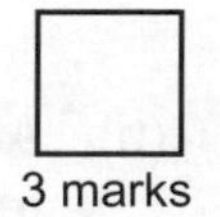

3 marks

(b) Calvin has a beaker with some of Sati's ink in it.
He uses the equipment shown below to separate out the water from the ink.

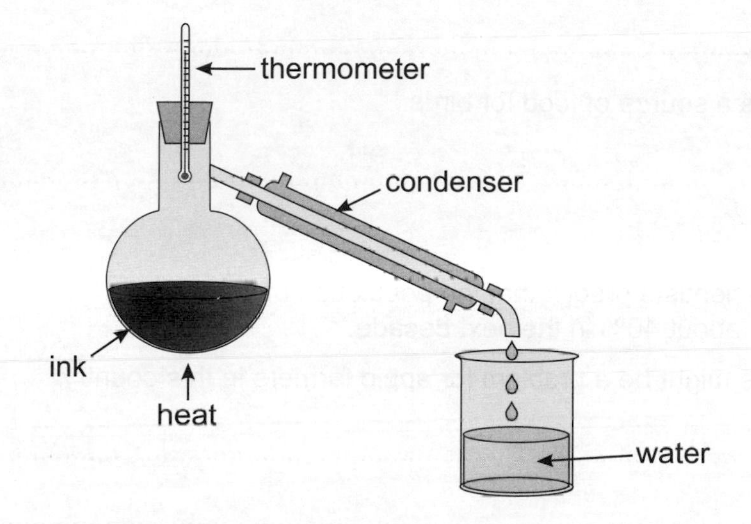

(i) Name the separation method that Calvin used.

....

1 mark

(ii) Calvin wants to know if the water he has collected in the beaker is pure.
Describe how Calvin could test whether the water is pure.

....

....

....

....

2 marks

8. Apple trees are pollinated by bees.

(a) Give **one** feature that you would expect apple blossoms to have, given that they are pollinated by bees. Explain your answer.

....

....

1 mark

(b) After an apple blossom is pollinated, apples begin to grow. The apple fruit helps the seeds travel away from the original tree before germinating and growing into new apple trees. Explain how each of the following characteristics helps the seed to move away from the original tree.

(i) The apple is round and heavy.

....

(ii) The apple is a source of food for birds.

....

2 marks

(c) In one country, scientists predict that the population of bees is likely to fall by about 40% in the next decade.

Suggest why this might be a problem for apple farmers in this country.

....

....

2 marks

9. A racing car is travelling around a circuit that is 6.3 km long.

(a) Over one lap of the track the average speed of the car was 52.5 m/s.
Calculate how long it took the car to complete the lap in minutes.

.... minutes

3 marks

(b) The horizontal forces that act on the racing car as it moves are the driving force and drag:

driving force

drag

Compare the sizes of the driving force and the drag when:

(i) The car is decelerating.

The driving force is ....

....

1 mark

(ii) The car is accelerating.

The driving force is ....

....

1 mark

(iii) The car is moving at a steady speed.

The driving force is ....

....

1 mark

10. Nitrogen gas (N_2) reacts with hydrogen gas (H_2) to make ammonia (NH_3).

(a) Ammonia is a molecule made up of nitrogen atoms and hydrogen atoms.

(i) What name is given to a substance made up of atoms of two or more different elements bonded together?

....

1 mark

(ii) Write down how many hydrogen atoms and how many nitrogen atoms are present in a molecule of ammonia.

Atoms of hydrogen: ....

Atoms of nitrogen: ....

1 mark

(b) The reaction between nitrogen and hydrogen is exothermic. What is an exothermic reaction?

....

....

1 mark

(c) Iron is often added to this reaction in industry, because it acts as a catalyst for the reaction.

What is a catalyst?

....

....

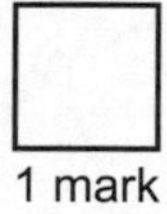

1 mark

11. A scientist is thinking about how to maintain the biodiversity of plant life on Earth in the future.

(a) What is meant by the term 'biodiversity'?

.... 1 mark

(b) The scientist wants to set up a plant gene bank. Describe what a plant gene bank is and how it can help to maintain biodiversity.

....

....

.... 2 marks

(c) Describe **one** other way of maintaining biodiversity.

....

.... 1 mark

12. The diagram below shows the human digestive system.

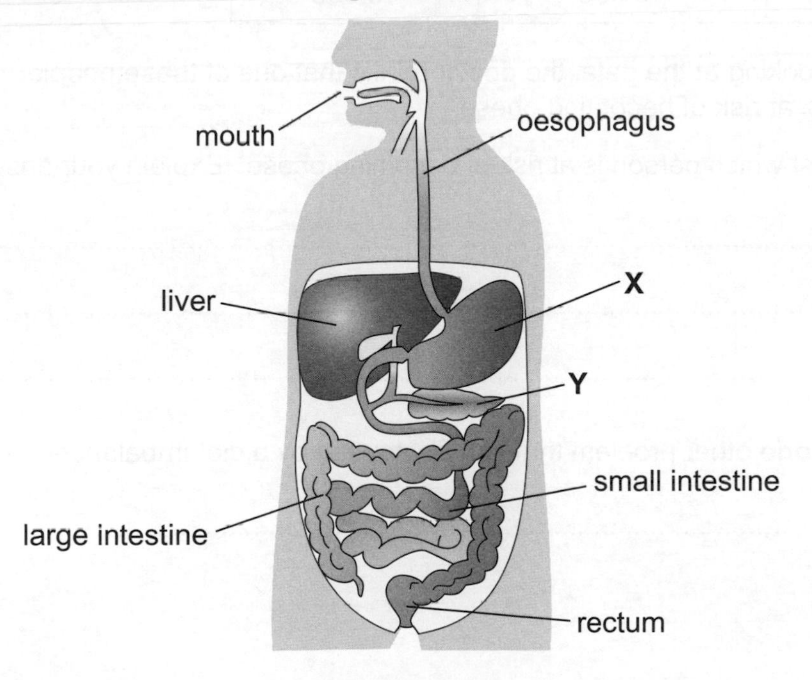

(a) (i) Name the parts of the digestive system labelled **X** and **Y**.

X is the ....

Y is the .... 2 marks

(ii) Name the part of the digestive system where bile is made.

....

1 mark

(iii) State the function of the large intestine.

....

1 mark

(b) A doctor is studying the diets and activity levels of five people.
The table below shows the daily basic energy requirement of each person.
It also shows their average daily energy intake and the extra energy used in activities each day.

Name	Daily basic energy requirement (kJ/day)	Average daily energy intake (kJ/day)	Average extra energy used for activities daily (kJ/day)
Amy	8400	8350	400
Declan	10 500	12 450	200
Maya	7800	7900	200
Noah	9900	11 800	3400
Tamir	10 000	10 050	15

(i) From looking at the data, the doctor thinks that one of these people may be at risk of becoming obese.

Suggest which person is at risk of becoming obese. Explain your answer.

....

....

....

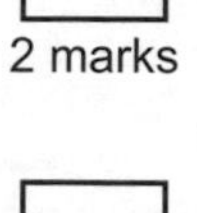

2 marks

(ii) Name **one** other problem that can be caused by a diet imbalance.

....

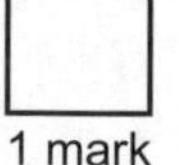

1 mark

13. Pete has 10 cm³ of sulfuric acid mixed in a beaker.
He adds sodium hydroxide solution to the beaker using a pipette.
Each time he adds 2 cm³ of sodium hydroxide, he records the pH of the mixture.

(a) Sulfuric acid reacts with sodium hydroxide to form a salt and water.
Name the salt formed by this reaction.

....

1 mark

(b) Pete drew the graph below to show the results of his experiment.

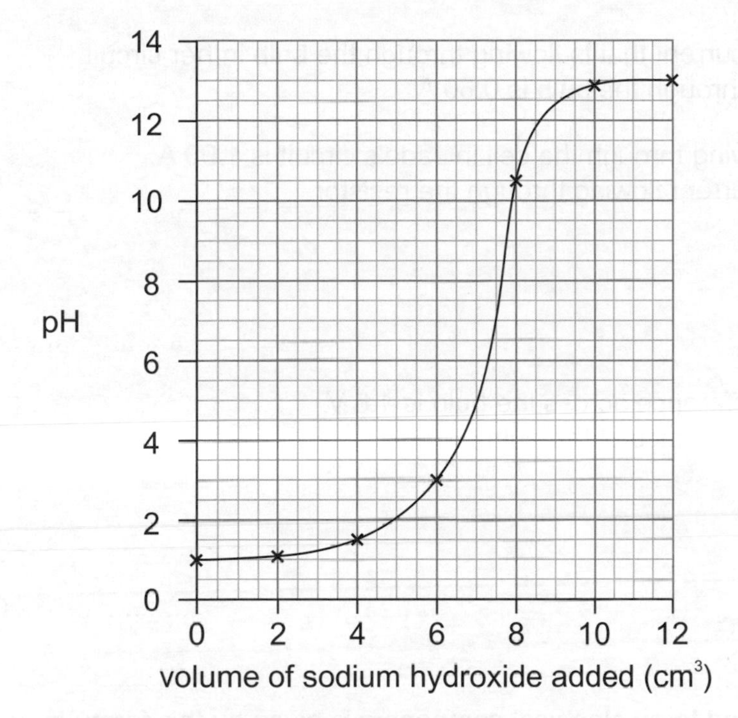

(i) Use the graph to find the pH of the mixture when Pete had added 6 cm³ of sodium hydroxide.

pH = ...

1 mark

(ii) Use the graph to estimate how much sodium hydroxide Pete would need to have added to make a neutral mixture.

.... cm³

1 mark

14. Jade has built a simple electrical circuit to power two components.
The circuit diagram for Jade's circuit is shown below.

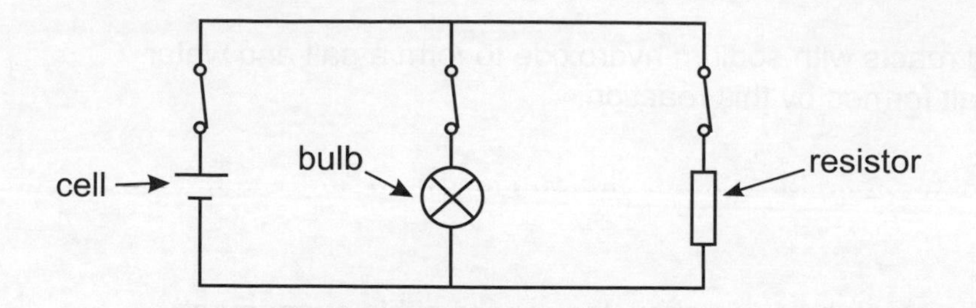

(a) Jade measures the current that is flowing through the bulb in her circuit.
The current flowing through the bulb is 0.60 A.

(i) The current flowing through the cell in Jade's circuit is 1.90 A
Calculate the current flowing through the resistor.

.... A

1 mark

(ii) The potential difference across the bulb is 1.5 V.
Calculate the resistance of the bulb.

.... Ω

2 marks

(b) The energy transferred to an electrical component is given by this formula:

energy (J) = power of appliance (W) × time (s)

(i) The bulb in Jade's circuit has a power rating of 40 W.
Calculate the energy, in joules, that the bulb transfers
when Jade leaves the circuit switched on for 6 minutes.

.... J

2 marks

(ii) Jade replaces the bulb with one with a much lower power rating.
Explain how this will this affect the energy transferred by the circuit
when it is switched on for 6 minutes?

....

....

2 marks

15. Mercury is the closest planet to the Sun.

(a) (i) Mercury is about 0.000006 light years from the Sun.
What is a light year?

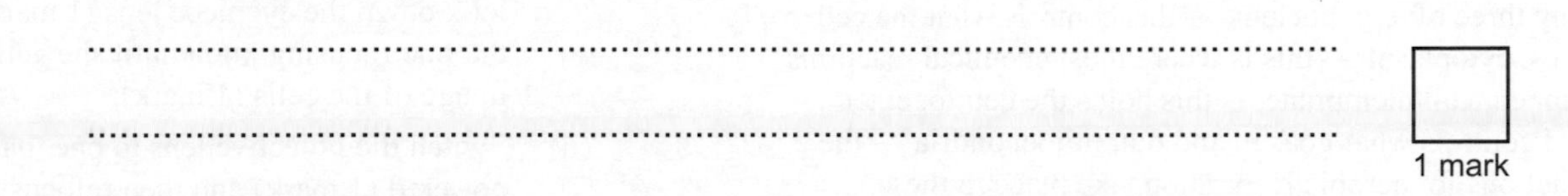

1 mark

(ii) How many seconds does it take light from the Sun to reach Mercury?
Give your answer to the nearest second.

.... s

2 marks

(b) An astronaut has a mass of 80 kg.

The gravitational field strength (g) on Mercury is 3.7 N/kg.
Calculate the weight of the astronaut if he was standing on Mercury.

.... N

2 marks

END OF TEST

Answers

Section One — Cells and Respiration

Page 5 — Warm-up Questions

1) To make it easier to see.
2) Any three of: e.g. nucleus — this controls what the cell does, cytoplasm — this is where most chemical reactions happen, cell membrane — this holds the cell together and controls what goes in and out, mitochondria — the reactions for aerobic respiration take place in these.
3) A tissue is made from a group of similar cells, while an organ is made from a group of different tissues working together.
4) Diffusion.
5) Aerobic respiration.
Glucose + oxygen → carbon dioxide + water (+ energy).
6) Aerobic respiration.

Pages 5-6 — Practice Questions

1 (a) (i) Nucleus (1 mark).
(ii) Cytoplasm (1 mark).
(iii) Mitochondria (1 mark).
(iv) Cell membrane (1 mark).
(b) Any two of: cell wall (1 mark), vacuole (1 mark), chloroplasts (1 mark).
(c) The cell wall gives support to the plant cell (1 mark).
(d) (i) Unicellular (1 mark).
(ii) E.g. an amoeba is a unicellular organism (1 mark). It is adapted to living in water and has a contractile vacuole (1 mark) to collect any excess water and squeeze it out at the cell membrane (1 mark).
OR: A euglena is a unicellular organism (1 mark). It is adapted to live in water and has a tail-like structure called a flagellum (1 mark) to help it swim (1 mark).

2 (a) The mitochondria (1 mark).
(b) (i) Anaerobic respiration (1 mark).
(ii) Glucose → lactic acid (+ energy) (1 mark).
(iii) E.g. during exercise (1 mark).

3 (a) **Cells** (1 mark) are the simplest building blocks of organisms. Several of these can come together to make up **a tissue** (1 mark) and several of these can work together to make **an organ** (1 mark).
(b) A group of organs which work together (1 mark).

4 (a) (i) The mirror (1 mark).
(ii) Direct sunlight (1 mark), because it can damage your eyes (1 mark).
(b) (i) She should select the lowest powered objective lens (1 mark). Then she should turn the rough focusing knob to move the objective lens down to just above the slide (1 mark). Next she should look down the eyepiece lens (1 mark) and adjust the fine focusing knob until she gets a clear image of the cells (1 mark).
(ii) Switch the objective lens to one that is higher powered (1 mark) and then refocus the microscope (1 mark).

Section Two — Humans as Organisms

Page 13 — Warm-up Questions

1) Carbohydrates and lipids (fats and oils)
2) E.g. vegetables/fruit/cereals
3) Obesity
4) The amount of energy needed to maintain your essential bodily functions each day.
5) Mechanically and chemically.
6) It means they speed up the rate of chemical reactions in the body.
7) The mouth
8) Big insoluble food molecules can't pass through the gut wall and be absorbed by the body. We need digestion to break the food down so we can absorb the nutrients it contains.
9) Bacteria

Page 13 — Practice Question

1 (a)

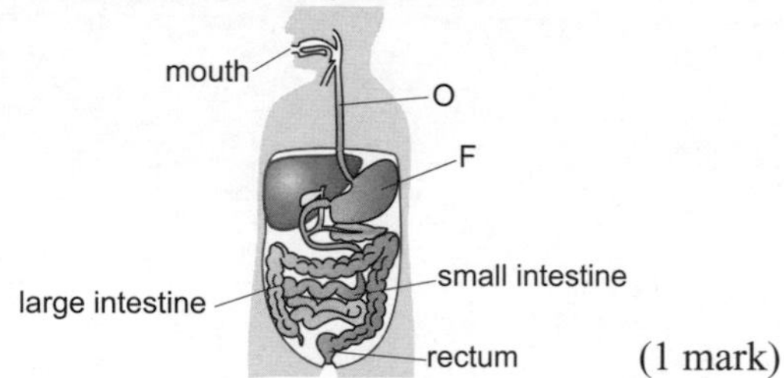

(1 mark)

(b) Any two from: digesting food (protein) / killing bacteria / churning up food.
(1 mark for each correct answer)
(c) They provide a large surface area for absorption (1 mark).
They have a good blood supply (1 mark).
They have a thin outer layer of cells (1 mark).
(d) The liver (1 mark).

Page 17 — Warm-up Questions

1) E.g. it is rigid/tough/strong
2) The skeleton provides a rigid frame that the soft tissues of the body are attached to.
3) Bone marrow
4) Tendons
5) Antagonistic muscles are pairs of muscles that work against each other. The muscles are attached to a bone. One muscle contracts and pulls on the bone while the other muscle relaxes, causing it to move (like a lever).
6) E.g. biceps and triceps / hamstrings and quadriceps.
7) Moment = force × perpendicular distance

Page 17 — Practice Question

1 (a) Support, protection, production of blood cells (1 mark for each).

(b) (i) Moment = force × perpendicular distance
= 15 × 0.35 = 5.25 Nm (1 mark)

(ii) 5.25 Nm (1 mark). To hold the arm still, the moment of the muscle must be the same as the moment caused by the weight of the box on the arm (but acting in the opposite direction) (1 mark).

If you've got the wrong answer in part (b)(i) but got the explanation right in (b)(ii), you'd still get all of the marks for (b)(ii).

(iii) Force = moment ÷ perpendicular distance
= 5.25 ÷ 0.05 = 105 N (1 mark)

Page 21 — Warm-up Questions

1) So that oxygen from the air can easily enter the bloodstream/body and carbon dioxide can easily leave the bloodstream/body.
2) The amount of air you can breathe into your lungs in a single breath.
3) Lung volume
4) E.g. any two of: the muscles you use to breathe will get stronger / your chest cavity will open up more so you will be able to get more air into your lungs / the number/size of blood vessels in your lungs can increase / the number of alveoli in your lungs can increase.
5) The muscles around their bronchioles contract, the lining becomes inflamed and fluid builds up.

Page 21 — Practice Questions

1 (a) Bronchus (1 mark) (b) Oxygen (1 mark)

(c) Carbon dioxide (1 mark)

(d) When you breathe in, the diaphragm contracts and moves down (1 mark) and the ribs move up (1 mark). This increases the volume of the chest cavity (1 mark), which decreases the pressure and causes air to rush into the lungs (1 mark).

2 (a) E.g. any two of: carbon monoxide / nicotine / particulates
(1 mark for each correct answer).

(b) E.g. tar from cigarette smoke damages the cilia in a smoker's airways (1 mark), which means they can't get rid of mucus properly (1 mark). The mucus sticks in the airways and causes them to cough (1 mark).

(c) E.g. bronchitis / emphysema / lung cancer / throat cancer / mouth cancer (1 mark).

Page 27 — Warm-up Questions

1) Sperm — they are made in the testes.
2) Day 14.
3) E.g. the placenta allows the blood of the baby to get close to the blood of the mother so that substances like oxygen, food and wastes can be exchanged.
4) A healthy mental state.
5) If the mother smokes, harmful chemicals from the cigarette smoke that enter her blood can cross the placenta and affect the foetus.
6) Recreational drugs are drugs that are used for fun.
7) E.g. any two from: paints / aerosols / glues.
8) E.g. any two from: lungs / brain / liver / kidneys.
9) E.g. any two from: ecstasy / LSD / solvents.

Pages 27-28 — Practice Questions

1 (a) Ovulation (1 mark)

(b) (i) The fallopian tubes / oviduct (1 mark)

(ii) When the nuclei of the egg and sperm join (1 mark)

(c) 39 weeks (1 mark)

2 (a)

name of organ	letter
sperm duct	B
glands	A
erectile tissue	C
scrotum	E
testis	D

(5 marks)

(b) Semen (1 mark)

3 (a) E.g. solvents (1 mark)

(b) Alcohol decreases the activity of the brain (1 mark).

(c) E.g. the brain and liver (1 mark for each correct answer).

(d) E.g. barbiturates (1 mark)

Page 29 — Section Two Revision Summary

7) BER = 5.4 × 24 × 54 = 6998.4 kJ/day

Section Three — Plants and Ecosystems

Page 34 — Warm-Up Questions

1) Light, carbon dioxide, chlorophyll and water.
2) Any three of: e.g. they're broad and have a large surface area / most of their chloroplasts are found in cells near the top of the leaf / the underside is covered in stomata / there are air spaces between leaf cells / they contain a network of veins.
3) The ovule
4) So the seeds can grow into new plants without too much competition from each other.

Pages 34-35 — Practice Questions

1 (a) Any three of: e.g. sunlight (1 mark), water (1 mark), carbon dioxide (1 mark), minerals/nutrients (1 mark).

(b) To attract insects for pollination (1 mark).

(c) They would hook onto the fur of animals (1 mark).

2 (a) carbon dioxide + water ⟶ glucose + oxygen (1 mark for each side correct.)

It doesn't matter which way round you get the words, as long as they're on the correct side of the equation.

(b) The plants under the tree may have received less water and sunlight (1 mark). This means they may not have been able to photosynthesise as much as the plants in the sunlight (1 mark).

(c) The group in mineral-rich compost (1 mark), because plants need minerals in order to stay healthy (1 mark).

3 (a) Any two from: e.g. the height the fruit is dropped from / where the experiment is done / the speed setting of the fan (1 mark for each correct answer).

(b) The sycamore fruit (1 mark), because its shape helps it to be dispersed by wind / it's lighter (1 mark).

4 (a) Jim is correct (1 mark).

(b)

stage	order
The nucleus of the male sex cell joins with the nucleus of the egg cell (ovule).	4 (1 mark)
Pollen grain lands on the stigma.	1
The ovary develops into a fruit with the seeds inside.	5 (1 mark)
The nucleus from a male sex cell moves down through the tube.	3 (1 mark)
A pollen tube grows down through the style to the ovary.	2 (1 mark)

Don't be thrown by all the fancy names involved in fertilisation — the process itself isn't that complicated really.

Page 38 — Top Tip Question

a) Fewer otters means more pike, which will eat more water beetles.

b) More pike would mean fewer perch, which could mean fewer water beetles get eaten.

Page 39 — Warm-Up Questions

1) They all depend on each other to survive.
2) Almost all energy on Earth comes from the Sun. Animals cannot make their own food from this energy. Therefore they rely on plants to capture and store the Sun's energy. This energy is then passed onto animals which eat plants and continues to be passed along food chains to other animals.
3) The arrows in a food chain or web show which organisms feed on each other / the direction of energy flow.
4) plankton → small fish → squid → whales
5) (a) A producer is an organism which uses the Sun's energy to 'produce' food energy.

(b) A carnivore is an organism that only eats animals, never plants.

(c) An omnivore is an organism that eats both plants and animals.

Page 39 — Practice Question

1 (a) (i) Any two from: e.g. the number of plants may (suddenly) decrease, the number of owls may (suddenly) increase, the number of jaegers may (suddenly) increase, the number of Arctic foxes may (suddenly) increase. (Maximum 2 marks.)

(ii) The number of Arctic foxes may also decrease (1 mark), as they have fewer food sources available (1 mark).

(b) Yes (1 mark) because more owls means fewer lemmings (1 mark). Therefore jaegers and Arctic foxes would become more dependent on geese as a food source (1 mark).

(c) Red foxes (1 mark).

(d) Toxic materials build up/accumulate as they are passed along a food chain (1 mark). The red foxes may be worst affected because they are at the top of the longest food chain shown (1 mark).

Owls are also at the top of a food chain but there are fewer stages between plants and owls than there are between plants and red foxes. Therefore you would expect red foxes to be the worst affected as more of the toxic material will have accumulated by the time it reaches the red fox.

Section Four — Inheritance, Variation and Survival

Page 45 — Warm-Up Questions

1) Chromosomes are long coiled up lengths of DNA.
2) The first model of DNA.
3) Yes — plants or animals with basically the same genes can have different characteristic features.
4) Characteristic features can be caused by genes (i.e. they're inherited) or the environment (surroundings).

5) The two classes of variation are continuous and discontinuous variation. Continuous variation means a feature can vary over a range of values, but features that vary discontinuously can only take certain values.
6) Scientists can store the genes of different species in gene banks so that scientists might be able to create new members of extinct species in the future. They can also prevent the destruction of habitats.
7) A habitat is an area where an organism lives. If habitats aren't preserved then the organisms will have nowhere to live, and they might die out.

Pages 45-46 — Practice Questions

1 (a) (i) DNA (molecule) (1 mark)
(ii) Gene (1 mark)
(iii) Chromosome (1 mark)
(b) (i) DNA is a long list of chemical instructions on how to build an organism (1 mark).
(ii) A double helix is a spiral of two chains wound together (1 mark).
(c) (i) Any three of, e.g. hair colour, eye colour, skin colour, height, hairiness etc. (1 mark for each correct answer).
(ii) 23 pairs (1 mark).
(iii) Heredity (1 mark).

2 (a) (i) Because there can be variation within a species (1 mark).
(ii) Discontinuous (1 mark) because the ears can only be straight or floppy (1 mark).
(b) E.g. rabbits with larger ears were better able to hear predators and avoid being eaten (1 mark). This meant rabbits with larger ears were more likely to survive and reproduce and pass on their big ear gene to their babies (1 mark). Over time the gene for big ears (and so the characteristic) became more common in the population (1 mark).

3 (a) Endangered means a species is at risk of becoming extinct (1 mark).
(b) Organisms are adapted to their environment, so if the environment changes they may become less well adapted (1 mark). So an organism may find it less easy to compete successfully for the resources it needs to survive and reproduce (1 mark). If this happens to the whole species it may become endangered (1 mark).
(c) Humans rely on plants and animals for food and to make things like clothing, medicines and fuel, so a species going extinct could impact on us (1 mark). There may be other uses for species we haven't identified yet (1 mark). Ecosystems are so complex that if one species becomes extinct, it could have a knock-on effect for other organisms (1 mark).
(d) (i) Seeds from endangered plants can be collected and stored in seed banks (1 mark). If the plants become extinct in the wild, new plants can be grown from the seeds kept in storage (1 mark).
(ii) Sperm and egg cells can be collected from animals, frozen and stored (1 mark). Scientists could then create new animal embryos from these cells in the future (1 mark).

Section Five — Classifying Materials

Page 53 — Warm-Up Questions

1) Gas.
2) Solid.
3) They move faster / their speed increases.
4) Changing state directly from solid to gas.
5) The pressure increases because the particles are squashed up in a smaller space, so they hit the sides of the container more often.

Pages 53-54 — Practice Questions

1 (a) Has a definite volume (1 mark). Has a high density (1 mark).
(b) E.g.

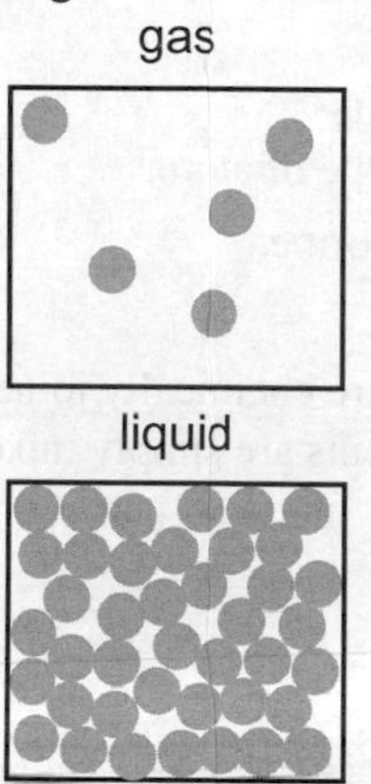

(1 mark for each.)

Don't forget that there should always be more liquid particles than gas ones in diagrams like this.

(c) (i) Diffusion (1 mark)
(ii) The deodorant gas particles move from areas of higher concentration to areas of lower concentration (1 mark) by bumping randomly into the air particles and spreading out as a result (1 mark).

2 (a) Air molecules exert pressure by colliding with the walls of the tyre (1 mark).
(b) The air molecules are compressed into a smaller volume because the car tyre gets squashed by the weight of the heavily loaded car (1 mark), so they hit the walls of the tyre more often (1 mark).
(c) (i) The air molecules will move faster (1 mark).
(ii) The molecules will collide with the wall of the tyre harder OR more frequently (1 mark).
(d) The tyre might blow out / explode / burst (1 mark).

This answer is just common sense really — you just have to imagine what might happen in real life if a tyre's pressure is too high.

3 (a) It has changed into a gas / it has evaporated (1 mark).
(b) Condensation (1 mark).

(c) (i) No — the particles themselves don't change, only their arrangement and their energy (1 mark).

(ii)

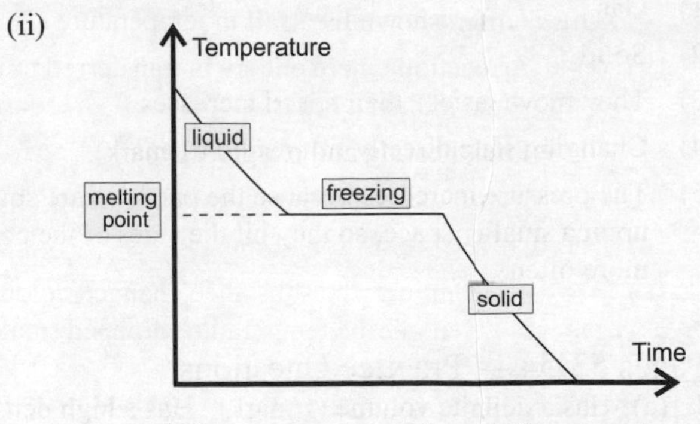

(1 mark for correct curve, 1 mark for correct labels)

Page 59 — Warm-Up Questions

1) An element contains only one type of atom.
2) Bromine is less reactive than fluorine.
3) Group 0
4) In a compound different atoms are chemically joined together. In a mixture the elements are simply mixed together, not chemically joined.
5) Iron sulfate

Page 59 — Practice Questions

1 (a) B (1 mark)

(b) A (1 mark)

Careful here — C is a mixture of an element and a compound, not a mixture of two compounds.

(c) B (1 mark)

(d) D (1 mark)

H_2O has two atoms of hydrogen and 1 atom of oxygen in each molecule. D is the only diagram that shows 2 of one type of atom and 1 of another.

2 (a) Copper oxide (1 mark)

(b) Copper carbonate (1 mark)

Page 65 — Warm-Up Questions

1) Mixtures, unlike compounds, are not chemically joined up.
2) A solute is the solid being dissolved. A solvent is the liquid the solute is dissolving into.
3) The solubility decreases as temperature decreases.
4) Grind up the rock salt with a pestle and mortar, add to a beaker of water and stir to dissolve the salt, filter the sand out through filter paper in a funnel, then evaporate off the water in an evaporating dish to leave the salt behind.
5) Chromatography is a method used for separating mixtures of coloured substances, e.g. separating dyes in inks.
6) Fractional distillation

Pages 65-66 — Practice Questions

1 (a) (i) Filter paper (1 mark)

(ii) Funnel (1 mark)

(b) Filtration / filtering (1 mark)

(c) She could leave the dish in a warm place for the water to evaporate slowly (1 mark).

2 (a)

Scientific term	Substance
solute	salt (1 mark)
solvent	water (1 mark)
solution (1 mark)	salt water

(b) (i) A saturated solution (1 mark).

(ii) Sally will be able to dissolve <u>more</u> (1 mark) salt in the water at the higher temperature, because the particles <u>are moving faster</u> (1 mark).

(c) (i) Simple distillation is a method used for separating out a mixture of a liquid and a solid (1 mark). The liquid is heated and boils off (1 mark). It's then cooled, condensed and collected, leaving the solid behind (1 mark)

(ii) She could check that its boiling point is 100 °C (1 mark).

Page 72 — Warm-Up Questions

1) E.g. iron, nickel and cobalt.
2) Most non-metal elements are gases.
3) a) metals
   b) non-metals
   c) metals
   d) non-metals
4) E.g. any four from: insulators to slow down energy being transferred by heating, insulators to reduce energy being transferred electrically, flexible, low density, easily moulded.
5) Concrete is made from a mixture of sand and gravel embedded in cement. It can withstand high compression stresses so it can support heavy things like buildings.

Pages 72-73 — Practice Questions

1 (a) (i) Because metals are strong and tough (1 mark).

(ii) Because metals are shiny when polished (1 mark).

(iii) Because metals can conduct energy by heating (1 mark).

(b) (i) No — you'd expect sulfur to be a poor conductor of energy by heating because it is a non-metal (1 mark).

(ii) Sulfur can be found on the <u>right</u> (1 mark)-hand side of the periodic table. It has a <u>brittle</u> (1 mark) consistency and a <u>dull</u> (1 mark) yellow surface.

2 (a) (i) Na OR Fe OR Cu (1 mark)
(ii) C OR He (1 mark)
(iii) He (1 mark)
(iv) Fe (1 mark)

Don't be put off by questions that are based on the periodic table — you just need to look at the information given in the diagram and use it to answer the questions. E.g. in this question, it's more about recognising symbols than using the periodic table.

(b) The periodic table contains more metals than non-metals (1 mark).

3 (a) (i) A polymer (also accept a specific polymer e.g. PVC) (1 mark).
(ii) E.g. metals have are heavy for their size, metals are not usually flexible (1 mark).
(b) (i) E.g. energy is not transferred by heating quickly in ceramics (1 mark).
(ii) E.g. car brakes (1 mark), spark plugs (1 mark).
(c) (i) A composite material is a material made from two or more materials stuck together (1 mark).
(ii) It is very strong (1 mark) and has a low density (so it is buoyant) (1 mark).

Page 74 — Section Five Revision Summary

20) a) magnesium oxide b) calcium oxide
c) sodium chloride d) calcium carbonate
e) copper sulfate

21) a) sodium chloride b) magnesium chloride
c) magnesium carbonate

Section Six — Chemical Changes

Page 79 — Warm-Up Questions

1) 92 g + 142 g = 234 g

The total mass after a reaction is always the same as the total mass before the reaction.

2) E.g. a gas comes off, a solid is made, the colour changes.
3) Endothermic — it takes in energy when the substance breaks down.
4) Catalysts come out of a reaction the same as when they went in — they're not chemically changed or used up.
5) Any two from: e.g. catalysts are expensive, different catalysts are needed for different reactions, catalysts may need to be cleaned, catalysts can be poisoned by impurities.
6) $CH_4 + 2O_2 \longrightarrow CO_2 + 2H_2O$

It's a good idea to learn the formulae for common substances — like oxygen (O_2), carbon dioxide (CO_2) and water (H_2O).

Page 79 — Practice Questions

1 (a) In an endothermic reaction, energy is **taken in from** the surroundings (1 mark).
This is often shown by a **fall** in temperature (1 mark).
(b) (i) A reaction where energy is transferred to the surroundings (1 mark).
(ii) Fuel, heating and oxygen (1 mark).
(iii) iron + oxygen ⟶ iron oxide (rust) (1 mark)

2 (a) Thermal decomposition (1 mark).
(b) Any two from: e.g. the substance changed colour, a gas was given off, the temperature dropped (maximum 2 marks).
(c) $CuCO_3 \longrightarrow CuO + CO_2$
(1 mark for $CuCO_3$ on the left of the equation, 1 mark for $CuO + CO_2$ on the right)
(d) Catalysts increase the speed of a reaction, not the amount of product made (1 mark).

Page 82 — Warm-Up Questions

1) 0-14
2) Acids have a pH below 7, alkalis have a pH above 7.
3) An indicator is a dye which changes colour depending on the pH of the solution it's in.
4) Litmus paper only tells you if something is an acid or an alkali, but universal indicator tells you what pH it is.
5) Sulfuric acid.

Page 82 — Practice Questions

1 (a) (i) Any one of: lemon juice, wine, milk (1 mark).
(ii) Sodium hydroxide solution (1 mark).
(iii) Water (1 mark)
(b) Any one of: lemon juice, wine, milk (1 mark), because these are acids and sodium hydroxide solution is an alkali (1 mark).

2 (a) He could remove a small sample and check with, e.g. Universal Indicator paper (1 mark) that the pH is 7/neutral (1 mark).
(b) acid + alkali ⟶ salt + water (1 mark)
(c) Nitric acid (1 mark).

Page 88 — Warm-Up Questions

1) Carbon and hydrogen.
2) Carbon is less reactive than sodium.
3) Magnesium is above hydrogen in the reactivity series.
4) The potassium.
5) a) Nothing — zinc is less reactive than magnesium.
b) The zinc strip becomes coated in copper and the solution goes colourless — zinc is more reactive than copper and displaces it in the solution.

Pages 88-89 — Practice Questions

1 (a) Any one of: e.g. he wore safety goggles, he stood the Bunsen burner on a heatproof mat (1 mark).

(b) (i) copper oxide + zinc → copper + zinc oxide (1 mark)

(ii) Displacement (1 mark).

The displacement reaction in this experiment is a bit odd because it involves solids (most of the ones you might come across tend to involve metal salts in solution). But don't be put off — it follows the same rules.

(iii) Zinc is more reactive than copper (1 mark).

(c) Nothing (1 mark). Zinc is less reactive than aluminium, so it would not displace aluminium from its oxide (1 mark).

2 (a) In order, starting with the most reactive: lithium (1 mark), magnesium (1 mark), mercury (1 mark), silver (1 mark).

You need to use both columns to work this one out. Silver is the only one with no reaction in oxygen, so that's the least reactive. Neither mercury nor silver react with dilute acids, so mercury must be next. Magnesium and lithium both react with dilute acids, but lithium reacts more quickly, so that must be the most reactive.

(b) magnesium + hydrochloric acid → magnesium chloride + hydrogen (1 mark)

(c) Potassium oxide (1 mark) — it reacts with acids but not alkalis, so it must be alkaline (1 mark).

3 (a) Carbon is less reactive than aluminium (1 mark), so it cannot displace aluminium from its compounds/ores (1 mark).

(b) Any one of, e.g. potassium, sodium, calcium, magnesium (1 mark).

(c) Because calcium is very unstable/reactive (1 mark).

Page 90 — Section Six Revision Summary

8) endothermic

11) $S + O_2 \rightarrow SO_2$

12) $2Ca + O_2 \rightarrow 2CaO$

Section Seven — The Earth and The Atmosphere

Page 95 — Warm-Up Questions

1) Tectonic plates
2) Sedimentary rock
3) Metamorphic rock
4) Element, compound, mineral, rock.
5) Weathering, erosion, transportation, deposition, burial/compression/cementation, heat/pressure, melting, cooling, exposure.
6) E.g. Onion skin weathering — the Sun warms the surface of a rock by day and by night it cools down, causing the surface to expand and contract. Eventually it breaks away, like peeling an onion.
Freeze-thaw weathering — water in a crack in a rock freezes and expands, making the crack bigger. After freezing and thawing many times, bits break off.

Pages 95-96 — Practice Questions

1 (a) A — crust (1 mark)
B — mantle (1 mark)
C — core (1 mark)

(b) The core / C (1 mark)

2 (a) (i) Igneous rock (1 mark)

(ii) The magma at A cooled **slowly** (1 mark), forming an **intrusive** (1 mark) rock which has **large** (1 mark) crystals, e.g. **granite** (1 mark). The lava at C cooled **quickly** (1 mark), forming an **extrusive** (1 mark) rock which has **small** (1 mark) crystals, e.g. **basalt** (1 mark).

(b) The metamorphic rock was formed by heat from the magma (1 mark) and pressure from the mass of the rock above (1 mark).

The question asked you how this particular rock was formed, so you need to include information from the diagram in your answer.

(c) (i) Pressure from the weight of the layers above squashes the water out (1 mark) and the sediments are cemented together by minerals (1 mark).

(ii) The relative age of the rocks (1 mark).

Page 100 — Warm-Up Questions

1) E.g. plastics, fuels/energy.
2) Recycling materials saves energy, which we usually get from burning fossil fuels.
3) Photosynthesis removes carbon dioxide from the air.
4) Nitrogen
5) A greenhouse gas is a gas which traps energy from the Sun in the Earth's atmosphere, e.g. carbon dioxide.
6) Global warming may increase melting of polar ice caps, causing sea levels to rise.

Page 100 — Practice Questions

1 (a) E.g. recycling uses less of the Earth's (limited) resources (1 mark). It uses less energy (1 mark). It saves money (1 mark). It reduces the amount of rubbish sent to landfill sites (1 mark).

(b) 1. True
2. True
3. False — recycling isn't free, it just costs less than sending materials to landfill and making them again from scratch.

(1 mark for correctly identifying all three statements as true or false, 1 mark for a correct explanation of the false statement).

2 (a) Decomposers take in carbon by feeding on dead plant and animal remains/animal waste (1 mark). They return carbon dioxide to the air (1 mark) through respiration (1 mark).

(b) Green plants remove carbon dioxide from the air (1 mark) through photosynthesis (1 mark). Decomposers release carbon dioxide into the air (1 mark).

Green plants return carbon dioxide to the air through respiration, just like decomposers. But the question asked you how the role of green plants is different to that of decomposers — and the answer to that is photosynthesis.

Section Eight — Energy and Matter

Page 107 — Warm-Up Questions

1) Gravitational potential, kinetic, elastic potential, thermal, chemical, electrostatic and magnetic.
2) joules
3) Radiation is where hot objects transfer energy by invisible waves to the surroundings.
4) Heating and light waves.
5) Energy transferred = force × distance.
   4 × 6 = 24 J.

Pages 107-108 — Practice Questions

1 (a) The energy is stored inside the battery (1 mark) in its chemical (1 mark) energy store.

(b) 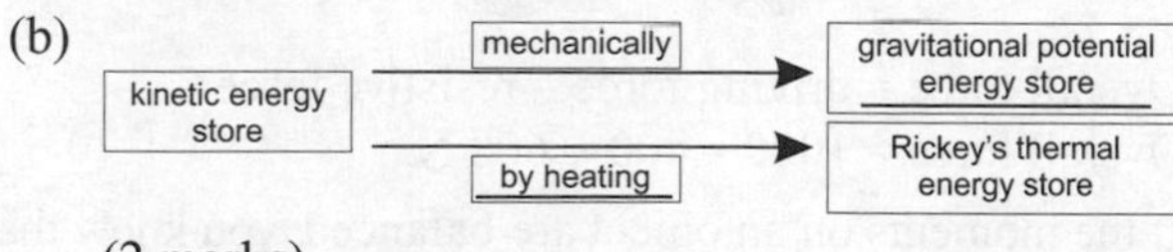

(2 marks)

2 (a) Energy in the **kinetic** energy store of the turbine is transferred away **electrically**. (1 mark)

(b) Energy in the **chemical** energy store of the coal is transferred by **heating** to the **thermal** energy stores of the room. (1 mark)

(c) Energy in the **gravitational potential** energy store of the skydiver is transferred **mechanically** to his **kinetic** energy store. (1 mark)

Remember — the seven possible energy stores are thermal, chemical, gravitational potential, kinetic, magnetic, electrostatic and elastic potential and the four ways of transferring are mechanically, electrically, by heating and by light or sound.

3 (a) (i) Her hand absorbs energy radiated by the hot cup (1 mark).

(ii) The cold cup radiates less energy than the hot cup (1 mark).

(b) (i) Conduction is energy transfer where vibrating particles pass on extra energy to neighbouring particles (1 mark).

(ii) Plastic is an insulator (1 mark). It slows down the rate at which the hot cup transfers energy to the table (1 mark).

Page 114 — Warm-Up Questions

1) Light is taken in by plants which use photosynthesis to transfer energy into the chemical energy stores of biomass.
2) The amount of electrical energy transferred (in kWh).
3) Cost = energy transferred (kWh) × price per kWh
4) 30 minutes = 1800 seconds
   energy transferred (J) = power (W) × time (s)
   60 × 1800 = 108 000 J (or 0.03 kWh).
5) The Sun
6) Any two from: e.g. save energy (turn lights off, drive cars with more fuel-efficient engines), recycle more, use more renewable energy resources.

Page 114 — Practice Questions

1 (a) A non-renewable energy resource is one that will run out (faster the more we use it). / Burning fossil fuels contributes to climate change / pollution. (1 mark)

(b) Any two from: e.g. wave, solar, biomass, wind, (1 mark for each correct answer, maximum 2 marks).

(c) It's important to use more renewable energy resources because non-renewables will run out (1 mark).

2 (a) Energy transferred (kWh) = power (kW) × time (h)
3 × 1.75 = 5.25 kWh (1 mark)

(b) Cost = energy transferred (kWh) × price per kWh
5.25 × 17 = 89.25p = £0.89, to the nearest penny (1 mark)

3 (a) The power rating of an appliance is the energy that it uses per second when it's operating at its recommended maximum power (i.e. when it's plugged into the mains) (1 mark)

(b) (i) 1 hour = 3600 s
Energy transferred = power × time
10 × 3600 = 36 000 J (or 0.01 kWh) (1 mark)

Be careful to convert the time to seconds if you're calculating the energy in joules.

(ii) The 60 W bulb transfers 60 × 3600 = 216 000 J (1 mark, also accept 0.06 kWh).
216 000 J — 36 000 J = 180 000 J
(1 mark, also accept 0.05 kWh).
OR
The power difference is 60 — 10 = 50 W
(1 mark).
50 × 3600 = 180 000 J
(1 mark, also accept 0.05 kWh).

Page 118 — Warm-Up Questions

1) 100 g (the mass does not change).
2) Sublimation is where a substance goes straight from being a solid to being a gas.
3) Brownian motion or the random movement of particles.
4) Any two from: e.g. solids are more dense, more difficult to compress, can't flow.

Page 118 — Practice Questions

1 (a) 136 g (1 mark)

(b) She could evaporate off the water to leave salt crystals (1 mark).

2 (a) Diffusion (1 mark).

(b) Particles bump and jiggle their way from an area of high concentration to an area of low concentration (1 mark). They constantly bump into each other, until they're evenly spread out throughout the gas (1 mark).

3 (a) The particles will move around more/speed up (1 mark).

(b) The size of the balloon will increase (1 mark) because the gas inside the balloon will expand (1 mark).

4 (a) (i) Solids are more dense than liquids (1 mark).

(ii) Water/ice (1 mark)

(b) E.g. solids can't flow, but liquids can (1 mark).

Page 119 — Section Eight Revision Summary

6) Energy transferred (J) = force × distance moved
= 2000 × 10 = 20 000 J = 20 kJ

19) Energy transferred = power (kW) × time (h)
= 1.5 kW × 0.5 h = 0.75 kWh

21) Cost = energy transferred in kWh × price per kWh
= 298.2 × 15 = 4473p = £44.73

23) The 300 W device (it has a higher power rating).

26) 50 g (the amount of substance is the same before and after).

Section Nine — Forces and Motion

Page 126 — Warm-Up Questions

1) speed = $\frac{\text{distance}}{\text{time}}$
The units are m/s, mph or km/h.
2) distance = speed × time = 5 × 30 = 150 m.
3) With a curved line / an increasing slope.
4) Your speed relative to the oncoming car is much higher than your speed relative to a car travelling in the same direction.
5) Unbalanced
6) Friction
7) The force of air resistance acting on the parachutist increases until it becomes equal to his weight, and so the forces on his body are balanced.

Pages 126-127 — Practice Questions

1 (a) The forces are balanced (1 mark).

If something isn't moving, then the forces acting on it must be balanced.

(b) Any two from:
1) It changes the ball's shape (squashes it) (1 mark).
2) It changes the ball's speed (1 mark).
3) It changes the ball's direction (1 mark).

(c) (i) Time = distance ÷ speed = 20 ÷ 50 = 0.4 s (1 mark)

Make sure that you're using the right units — in this case the answer should be in seconds.

(ii) Air resistance has slowed the ball down slightly (1 mark).

2 (a) (i) 10 – 5 = 5 m (1 mark)
(ii) speed = distance ÷ time
5 ÷ 10 = 0.5 m/s
(1 mark for correct answer, 1 mark for using the distance found in part (i)).

(b) 20 seconds (1 mark).

The model train is stationary for 10 seconds between 30 and 40 seconds, then another 10 seconds between 50 and 60 seconds.

(c) The train is moving back towards its starting point (1 mark).

(d) Any two from: e.g. air resistance/friction / Trevor applied the brakes / Trevor turned off the power (1 mark for each correct answer).

3 (a) 600 + 725 = 1325 km/h (1 mark).

The aircraft are travelling in the opposite direction to each other, so add their speeds.

(b) Subtract the speed of one aircraft from the speed of the other aircraft (1 mark).

Page 129

1) Balanced. 2) Balanced. 3) Unbalanced — right side down.
4) Unbalanced — left side down.
5) Balanced. 6) Balanced.

Page 133 — Warm-Up Questions

1) E.g.

2) Overall force = driving force – resistive force
Overall force = 1000 – 400 = 600 N.
3) If the moments on an object are balanced, you know that:
<u>anticlockwise</u> moments = <u>clockwise</u> moments.

You could write anticlockwise and clockwise the other way round here.

4) It means all the forces acting on the spring are balanced.
5) pressure = $\frac{500}{0.002}$ = 250 000 N/m^2 or Pa
6) The upthrust is from the water is less than the object's weight.

Pages 133-134 — Practice Questions

1 (a) Elastic potential energy store (1 mark)
(b) Force = spring constant × extension
= 100 × 0.2 = 20 N (1 mark)
(c) 20 N (1 mark)

This should be the same as your answer to (b).

2 (a) 8 cm = 0.08 m
Moment = force × perpendicular distance.
Moment = 6 × 0.08 = 0.48 (1 mark) Nm (1 mark).

Always include the units in number questions. You can lose marks if you forget the units.

(b) force = $\frac{\text{moment}}{\text{distance}} = \frac{0.48}{0.005}$ = 96 N (1 mark).
(c) pressure = $\frac{\text{force}}{\text{area}} = \frac{96}{0.00001}$ = 9 600 000 (1 mark) N/m^2 (1 mark).

3 (a) Martin's mother's weight is spread over a much smaller area when her weight's on her heels (1 mark) than Martin's so the pressure is higher (1 mark). (Answers stating that Martin's feet have a larger surface area so the pressure is lower are also acceptable.)
(b) Pressure = force / area (1 mark)
area = 0.0001 m^2 × 2 = 0.0002 m^2 (1 mark)
Pressure = 600 N / 0.0002 m^2
= 3 000 000 N/m^2 or Pa (1 mark)

You've been asked to give the pressure when she's standing on <u>both</u> feet, so you have to double the area of one of her shoes to get the right answer.

4 (a) The pressure felt by the dolphin will increase (1 mark).

(b) (i) Upthrust causes the bubble to rise (1 mark).

(ii) E.g.

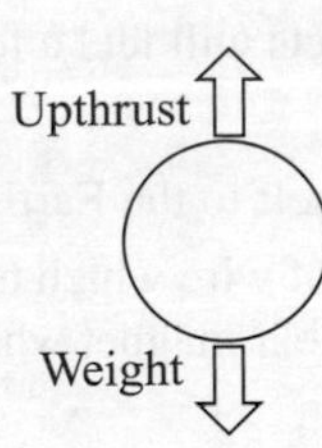

(1 mark for correct forces, 1 mark for each arrow being the same size by eye.)

The bubble is rising at a steady speed, so the forces are balanced. Make sure the arrows are the same size.

Page 135 — Section Nine Revision Summary

3) s = d/t = 5 ÷ 2 = 2.5 m/s.

4) s = d/t = 100 ÷ 20 = 5 m/s. Yes.

5) s = d/t so d = s × t = 40 × 0.25 = 10 miles. (15 minutes = 0.25 hours)

25) 50 × d = 100 so d = 2 m.

29) 200/2 = 100 N/m^2 (or 100 Pa).

Section Ten — Waves

Page 143 — Warm-Up Questions

1) Water waves are transverse waves.

2) False

3) The maximum displacement of a wave.

4) A convex lens bulges outwards, so it causes rays of light to converge (move together) to a focus.

5) Red light has the lowest frequency.

6) E.g. a prism.

Pages 143-144 — Practice Questions

1 (a)

Tyrone's clothes	colour in red light	colour in green light
green T-shirt	Black	Green
blue jeans	Black	Black
red boots	Red	Black
white laces	Red	Green

(4 marks available, 1 mark for each two correct answers.)

You might be asked about different situations, but don't get confused — all you need to do is apply the rules you've learnt in this section to the question.

(b) (i) Blue (1 mark).

(ii) Because the blue light is reflected and all the other colours are absorbed (1 mark).

2 (a) (i) The cornea does most of the eye's focusing (1 mark).

(ii) The lens changes shape to focus light from objects at varying distances (1 mark).

(iii) The iris controls the amount of light entering the eye (1 mark).

(b) (i) Cells in the retina are photo-sensitive/ sensitive to light (1 mark). When light waves hit a retina cell, they cause chemical and electrical changes in these cells that send signals to the brain (1 mark).

(ii) E.g. the digital image sensor (1 mark).

3 (a) Paper has a rough surface (1 mark). So light is reflected back (scattered) in lots of different directions (1 mark).

Remember, this is called diffuse scattering or diffuse reflection.

(b) (i) Specular reflection (1 mark) (ii) 39° (1 mark)

Remember, angle of incidence = angle of reflection.

(c) (i) Towards the normal (1 mark).

(ii) E.g.

(1 mark for light hitting the glass at an angle to the normal, 1 mark for light bending away from the normal as it leaves the glass block.)

(d) (i) Light waves are transverse waves (1 mark).

Light waves are the same type of wave as water waves.

(ii) Light travels at 3×10^8 m/s (also accept 300 000 000 m/s or three hundred million metres per second) (1 mark).

(iii) It slows down / the speed decreases (1 mark).

4 (a) (i) The displacements combine, so that the crest height doubles (1 mark).

(ii) The displacements combine, so that the two waves cancel each other out (1 mark).

(b) Superposition (1 mark).

Page 149 — Warm-Up Questions

1) Light.

2) A vacuum.

3) The frequency of the sound / the number of vibrations per second.

4) 20 Hz.

5) E.g. information.

6) E.g. carpets, curtains

7) The diaphragm.

Page 149 — Practice Questions

1 (a) It vibrates (1 mark).

(b) The cochlea (1 mark).

(c) (i) Ultrasound (1 mark).

(ii) Ultrasound has a higher frequency/pitch than the normal auditory range of humans (1 mark).

(iii) E.g. cleaning objects such as jewellery, false teeth or fountain pen nibs (1 mark).

2 (a) The sound travels faster through the floor than it does through the air (1 mark).

(b) The sound is reflecting off the walls of the school hall / he's hearing an echo of the sound (1 mark).

Section Eleven — Electricity and Magnetism

Page 156 — Warm-Up Questions

1) Potential difference
2) Ohms (Ω)
3) An insulator is a material that doesn't easily allow electric charges to pass through it, e.g. wood, plastic, rubber, glass, ceramics (other answers possible).
4) (a) − + or + − (b) E.g. (c)
5) Ammeters measure current, voltmeters measure potential difference.
6) A battery rating tells you the potential difference it will supply. A bulb rating tells you the maximum potential difference that you can safely put across it.
7) A parallel circuit is more useful than a series circuit because the devices in it can be operated independently.

Page 156 — Practice Question

1 (a) Parallel (1 mark).

(b) (i) It would stay on (1 mark).

(ii) It would go off (1 mark).

(c) Bulb A will stay on because the electricity is still flowing through it / the circuit is still complete (1 mark). Bulb B will go off because there is no electricity flowing through it / the circuit is broken (1 mark) because wood does not conduct electricity well (1 mark).

(d) (i) 0.10 + 0.02 = **0.12 A** (1 mark)

Remember the total current in a parallel circuit is the same as the current through each branch added together.

(ii) Current = 0.02 A (1 mark)
Resistance = potential difference ÷ current
= 0.1 ÷ 0.02 = **5 Ω**
(1 mark for the correct answer, plus 1 mark for the correct units)

The current in the copper strip is the same as the current in the bulb on the same branch as the strip.

Page 160 — Warm-Up Questions

1) Static charges are caused by the transfer of electrons.
2) An electric field is the space around a charged object where other charged objects will feel a force.
3) E.g. iron.
4) The magnet is aligning itself to the Earth's magnetic field.
5) A solenoid is a long coil of wire which has a magnetic field that is the same as a bar magnet when a current flows through the wire.
6) A simple electric motor consists of a loop of coiled wire in a magnetic field.

Page 160 — Practice Questions

1

The field lines should point away from the north pole(s) and towards the south pole(s).
(2 marks available — 1 mark for correct diagram and 1 mark for correct explanation.)

2 (a) Any one of: increase the number of coils in the wire (1 mark), increase the current (by adding another cell) (1 mark).

(b) The compass will align itself to the electromagnet's magnetic field (1 mark), pointing directly towards/ away from the iron bar (1 mark).

Remember — a compass needle always points from N to S along the field lines of a magnetic field. At the end of a solenoid, the field lines point directly into or out of the solenoid. You can't tell which end is north and which is south here, so you get the mark whether you've said it points into or out of the iron bar.

Page 161 — Section Eleven Revision Summary

7) Component C (it has the smallest resistance).

14) 3 A

Section Twelve — The Earth and Beyond

Page 165 — Warm-Up Questions

1) (The force due to) gravity.
2) One (the Sun).
3) E.g. stars give out light, but planets don't.

Page 165 — Practice Questions

1 (a) B (1 mark). As the Earth spins, part of it will face the Sun, which is day, and then turn away from it, which is night (1 mark).

(b) A year (1 mark).

(c) 24 hours (or 1 day) (1 mark)

(d) 365 ¼ (1 mark).

2 (a) D (1 mark).

(b) B (1 mark).

3 (a) The strength of gravity on Mars is weaker than it is on Earth (1 mark).

(b) weight = mass × gravitational field strength (1 mark).
500 × 3.7 = 1850 N (1 mark).

Page 166 — Section Twelve Revision Summary

8) weight = mass × gravitational field strength
$5 \times 25 = 125$ N

Section Thirteen — Exam Practice

Page 167 — Test 1

1 C
2 gravitational potential (energy store)
3 tissue
4 D
5 $F = k \times e$ / The extension of an object is directly proportional to the force applied.
6 Any two from: movement / reproduction / sensitivity / nutrition / excretion / respiration / growth.
7 A
8 E.g. in a physical change, no chemical reaction takes place and no new substances are made, unlike in a chemical change.
9 Any two from: e.g. stiff / brittle / good thermal insulator / good electrical insulator.
10 The thin, outer layer of solid rock (that we live on).

Page 168 — Test 2

1 Any two from: e.g. it uses less resources than creating a new item / it uses less energy than creating a new item / it is cheaper than creating a new item / it means less rubbish is sent to landfills.
2 Conduction and radiation
3 A
4 E.g. biceps and triceps / hamstrings and quadriceps
5 A
6 Protection, support, production of blood cells and movement.
7 C
8 B
9 C
The total current through the circuit is the sum of the currents through each branch, i.e. total current = 5 A + 10 A = 15 A.
10 $3S + 2O_3 \rightarrow \mathbf{3}SO_2$

Page 169 — Test 3

1 B
2 It changes from a solid to a gas.
3 B
4 C
5 6 Ω
6 Thermal decomposition is when a substance breaks down (into at least two other substances) when it is heated.
7 All the forces acting on the object are balanced / There is a resultant force of zero on the object.
8 C
9 A biological catalyst that speeds up the rate of chemical reactions.
10 B

Page 170 — Test 4

1 An ecosystem is all the living organisms in one area, plus their environment.
2 A
3 E.g. metals are generally good conductors of electricity and non-metals generally aren't / metals are generally good thermal conductors but non-metals are generally thermal insulators / metals are generally strong and hard-wearing whereas non-metals generally aren't / non-metals are generally dull but metals can be polished to a shine / metals are generally ductile whilst non-metals are generally brittle / metals are dense whilst non-metals generally have low densities / metals generally have high melting and boiling points, whereas non-metals generally don't / some metals are magnetic, but non-metals are not magnetic.
4 Electromagnets are magnets whose fields can be turned on and off by an electric current / are magnets that are made from current-carrying wires.
5 B
6 3×10^8 m/s
7 A
8 inherited/passed down from parents
9 A
10 D

Page 171 — Test 5

1 Something that eats both plants and animals.
2 C
3 B
4 Potassium, magnesium, zinc, copper.
5 20-20 000 Hz
6 C
7 Discontinuous variation means a feature can only take certain values.
A person's blood group is an example of discontinuous variation — it can only take the values A, B, AB or O.
8 D
Remember to convert the time to seconds, before using energy transferred = power × time.
9 B
10 Calcium oxide.

Page 172 — Test 6

1 A

2 A sex cell

3 B

4 Energy is taken in from the surroundings.

5 C

6 No

Objects float when the upthrust on them equals their weight. If the weight of an object is larger than its upthrust, the object will sink.

7 D

8 Any two from: e.g. decreases brain activity/slows responses / impairs judgement / damages the liver/can cause cirrhosis.

9 D

10 Respiration and combustion

Pages 173-187 — Practice Exam

1 a) W ***[1 mark]***

b) i) Z ***[1 mark]***

ii) Any two from: e.g. the element will conduct electricity/will be an electrical conductor / will be a good conductor of energy by heating/will be a thermal conductor / will be strong/tough / will be shiny / will be ductile / will have a high melting point / will have a high boiling point / will have a high density
[2 marks — 1 mark for each correct property]

2 a) i) B ***[1 mark]***

ii) C ***[1 mark]***

iii) A ***[1 mark]***

b) i) **Sedimentary** rocks are formed from layers of rock fragments and dead matter, laid down and squashed together over many years ***[1 mark]***.

ii) **Igneous** rocks have no layers. They are made up of lots of crystals, which may be large or small ***[1 mark]***.

3 a) i)

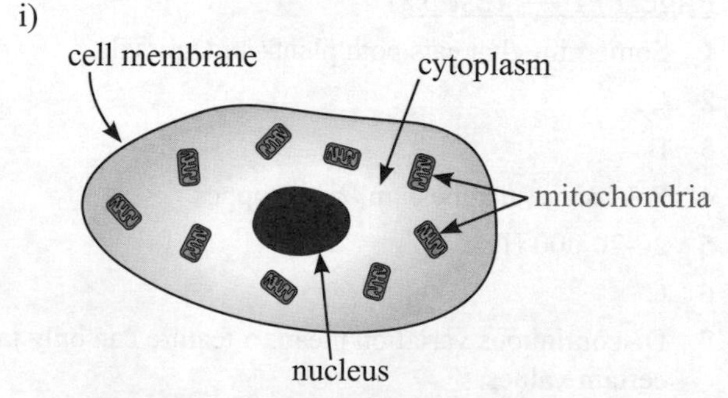

[3 marks — 1 mark for each correct label]

ii) They are where most of the reactions for aerobic respiration take place ***[1 mark]***.

b) To allow the organism to move about / swim ***[1 mark]***.

4 a) **glucose** + oxygen → **carbon dioxide** + water (+ energy)
[1 mark for both correct]

b) i) Any two from: e.g. they are moist / they have a good blood supply / they have a big inner surface area ***[2 marks — 1 mark for each correct feature]***

ii) E.g. tar covers the cilia in the lungs which damages them ***[1 mark]***

c) It is respiration without oxygen ***[1 mark]***.

5 a) i) The smoke particles were being moved about by collisions with air particles travelling at high speeds ***[1 mark]***.

(ii) Brownian motion ***[1 mark]***

b) The smoke particles diffused ***[1 mark]*** from an area of high concentration (where the smoke was released) to an area of low concentration (near Jack) ***[1 mark]***.

6 a)

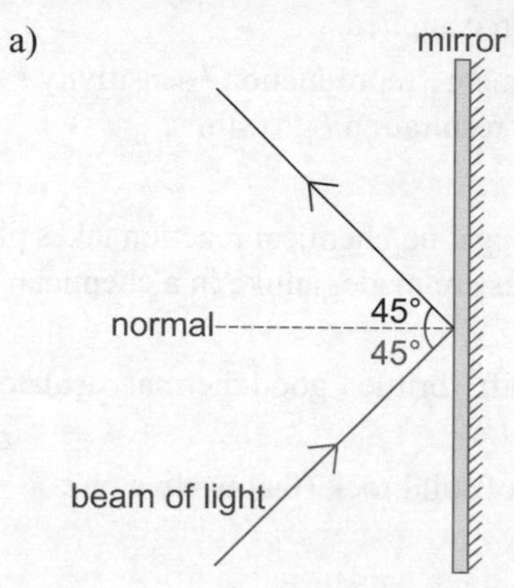

[1 mark for a 45° angle between the normal and the reflected beam, 1 mark for an arrow on the reflected beam to show that it's moving away from the mirror.]

The angle doesn't have to be labelled to get the first mark here, as long as it's 45°.

b) Specular ***[1 mark]***.
The mirror is smooth, so all the light reflects at the same angle ***[1 mark]***.

c) retina ***[1 mark]***

7 a) i) Solvent: water ***[1 mark]***

ii) Solution: ink ***[1 mark]***

iii) Solute: (blue) dye ***[1 mark]***

b) i) (simple) distillation ***[1 mark]***

ii) Calvin can measure the boiling point of the water he has collected ***[1 mark]***. If the water is pure, it will boil at exactly 100 °C / the boiling point of pure water ***[1 mark]***.

8 a) E.g. brightly coloured petals/scented flowers to attract the bees / nectaries to produce nectar to feed/attract the bees / sticky stigma to take the pollen off the bees ***[1 mark for any correct feature with a sensible explanation.]***

b) (i) The apple will fall from the tree and roll away from it ***[1 mark]***.

(ii) The apple will be eaten by birds, and the seeds will eventually be released in their faeces away from the original tree ***[1 mark]***.

c) E.g. if there are fewer bees, fewer apple flowers will be pollinated ***[1 mark]***, so the trees will produce less fruit ***[1 mark]***.

9 a) speed = 52.5 m/s
distance = 6.3 km = 6.3 × 1000 = 6300 m *[1 mark]*
time = distance ÷ speed
= 6300 ÷ 52.5
= 120 s *[1 mark]*
120 s = 120 ÷ 60 = 2 minutes *[1 mark]*

You'd also get the first mark here if you converted the speed from m/s to km/s and used that with the distance in km. The important thing is that all the units match.

b) i) The driving force is **less than the drag** *[1 mark]*.
ii) The driving force is **greater than the drag** *[1 mark]*.
iii) The driving force is **equal to the drag** *[1 mark]*.

10 a) i) a compound *[1 mark]*
ii) Atoms of hydrogen: 3
Atoms of nitrogen: 1
[1 mark for both correct]
b) A reaction which transfers energy to the surroundings *[1 mark]*.
c) A substance which speeds up a chemical reaction *[1 mark]*.

11 a) The variety of different species living in an area *[1 mark]*.
b) A plant gene bank is a collection of seeds that are stored instead of grown *[1 mark]*. If any of the plants become extinct in the wild, new plants can be grown from the stored seeds *[1 mark]*.
c) E.g. prevent destruction of habitats that could cause extinction of species *[1 mark]*.

12 a) i) X is the **stomach** *[1 mark]*.
Y is the **pancreas** *[1 mark]*.
ii) liver *[1 mark]*
iii) The large intestine absorbs water from food and drink *[1 mark]*.
b) i) Declan *[1 mark]*, because he is taking in more energy from his diet than his daily basic requirement, and he is not using it up with extra activities *[1 mark]*.

Your body stores excess energy from your diet as fat. So if you take in a lot more energy than you use up, you will put on a lot of weight.

ii) E.g. starvation / deficiency diseases *[1 mark]*

13 a) sodium sulfate *[1 mark]*.
b) i) 3 *[1 mark]*
ii) 7.5 cm^3 *[1 mark for an answer between 7.2 cm^3 and 7.7 cm^3]*

A neutral mixture has a pH of 7.

14 a) i) current through component X
= current through cell – current through bulb
= 1.90 A – 0.60 A = 1.3 A *[1 mark]*
ii) resistance = potential difference ÷ current
= 1.5 ÷ 0.60 *[1 mark]*
= 2.5 Ω *[1 mark]*
b) i) 6 minutes = 6 × 60 = 360 seconds *[1 mark]*
energy = power of appliance × time
= 40 × 360
= 14 400 J *[1 mark]*
ii) The energy transferred will be lower, because the power rating is lower *[1 mark]*. So less energy will be transferred each second *[1 mark]*.

15 a) i) A unit of distance equal to how far light travels in one year *[1 mark]*.
ii) Mercury is 0.000006 light years, so it takes 0.000006 years for light to reach Mercury from the Sun. In seconds:
time = 0.000006 × 365 × 24 × 60 × 60 *[1 mark]*
= 189.216 s = 189 s *[1 mark]*
b) Weight = mass × gravitational field strength
= 80 × 3.7 *[1 mark]*
= 296 N *[1 mark]*

Index

Index

Index